Medical Internet of Things

Medical Internet of Things
Techniques, Practices and Applications

Edited by
Anirban Mitra
Jayanta Mondal
Anirban Das

CRC Press
Taylor & Francis Group
Boca Raton London New York

CRC Press is an imprint of the
Taylor & Francis Group, an **informa** business

A CHAPMAN & HALL BOOK

First Edition published 2022
by CRC Press
6000 Broken Sound Parkway NW, Suite 300, Boca Raton, FL 33487-2742

and by CRC Press
2 Park Square, Milton Park, Abingdon, Oxon, OX14 4RN

© 2022 Taylor & Francis Group, LLC

CRC Press is an imprint of Taylor & Francis Group, LLC

ISBN: 978-0-367-33123-8 (hbk)
ISBN: 978-1-032-12493-3 (pbk)
ISBN: 978-0-429-31807-8 (ebk)

DOI: 10.1201/9780429318078

Typeset in Palatino
by MPS Limited, Dehradun

Contents

Preface

In recent times, Medical Internet of Things (MIoT) has emerged as one of the most helpful technological gifts to mankind. With the incredible development in data sciences, big data technologies, IoT and embedded systems, today it is possible to collect a huge amount of sensitive and personal data, compile it and store it through cloud or edge computing techniques. However, important concerns remain about security and privacy, preservation of sensitive and personal data, efficient transfer, storage and processing of MIoT-based data.

This book is an attempt to explore new ideas and novel techniques in the area of MIoT. The book is composed of fifteen chapters discussing the basic concepts, issues, challenges, case studies and applications in MIoT.

Chapter 1 is based on applications of MIoT in the healthcare industry. This chapter describes sensor-based devices like smartphones that sense and capture real-time health domain data and analyze the data. Chapter 2 explores a brain-computer interface. The authors discuss a system that can serve patients with speech impairment, paralysis or other motor disabilities. Chapter 3 describes medical image preprocessing techniques and elaborates on applications for different microscopic cytopathological images.

Chapter 4 is about the Internet of Vehicles (IoV) and provides an overview on its use in the medical sector. The chapter discusses the healthcare-related system associated with IoV that can be used for sensing and processing real-time data for efficient decision making. Chapter 5 is on healthcare-related image classification for accurate diagnosis of complex medical disease. Specifically, it discusses processing images associated with melanoma versus visual diagnosis.

Chapter 6 focuses on concepts of IoT and fog computing used for medicinal plant irrigation. It elaborates on irrigation with IoT, fog and related computing-based automation tools. Chapter 7 discusses IoT networks in the healthcare domain and how they can provide constant and better quality healthcare-related services to patients. Chapter 8 describes the basic techniques, practices and applications of MIoT and focuses on various medical devices and applications that connect to healthcare IoT systems through online computer networks that serve the healthcare industry.

Chapter 9 provides a case-based analysis of MIoT, including the basic architecture that forms the framework for supporting healthcare-based IoT applications. Chapter 10 describes an application of MIoT-based infrastructure for the treatment of cardiac patients using various wearable sensors and devices as well as for efficiently storing and transmitting patients' data.

Chapter 11 focuses on issues related to paralyzed patient healthcare. Interconnected devices are being used to gather data, monitor patients and make decisions in order to provide patients with better healthcare. An application of MIoT is elaborated on in Chapter 12 The system relates to a small electronic device that can be used to guide a person with visual impairment or disabilities within a building, with or without prior knowledge, and also in outdoor environments such as fields or roads. Concepts and issues related to detection and analysis of human emotions using MIoT techniques is discussed in Chapter 13

Chapter 14 provides an introduction to the role of explainable artificial intelligence (XAI) in Internet of Things (IoT) based disease prediction and diagnosis. Chapter 15 provides an overview of sleep stages and sleep behavior. The main objective of this chapter is to analyze the sleep abnormalities shown in different transition states of sleep. This chapter can help clinicians and researchers to design the most cost-effective and user-friendly systems to diagnose sleep disorders.

Our intention in editing this book was to offer novel advances and applications of MIoT in a precise and clear manner to the research community to provide in-depth knowledge in the field. We hope this book will help those interested in this field as well as researchers to gain insight into different concepts and their importance in multifaceted applications of real life. This has been done to make the edited book more flexible and to stimulate further interest in the topic.

Anirban Mitra
Jayanta Mondal
Anirban Das

Editors

Dr. Anirban Mitra (PhD in Science & Technology - Computer Science) is currently working as an Associate Professor, Research & NTCC (Non-Technical Credit Course) Coordinator in the Department of Computer Science and Engineering (Amity School of Engineering and Technology) of Amity University Kolkata and is also associated with 'AUK-IT and Tech club' (a University level IT Club) as Co-Guide. He has 13+ Years of experience in academics. Apart from his regular teaching, mentoring and research guidance, he is actively involved with the academic and research world as an author of books, book chapters and a number of papers in reputed journals and conferences. He is a senior member of IEEE and ACM and member of other reputed societies like CSI (India). He is associated with several journals, conferences and technical workshops as guest editor, reviewer, technical committee member, member of publicity committee, session chair and as an invited speaker. He has played an important role (including delivering keynote talks, invited talks) in participating and organizing several successful academic and research-oriented events. Apart from research guidance and supervising PhD Scholars, his name is included in the list of PhD examiners at several Universities. He was associated with a funded research project, copyrights, and patent. His research area includes Data Science and its application in areas of Knowledge Representation, Social Computing, Social Networking, and Graph Theory.

Dr. Jayanta Mondal is currently working as an Assistant Professor in the School of Computer Engineering, KIIT (Deemed to be) University, Bhunaneswar. Previously, he has worked in reputed universities like MODY University, Laxmangarh and UEM, Kolkata. He completed his PhD in Computer Science and Engineering in 2018 from KIIT as an Institute Scholar. He has over nine years of experience, including three years after his research degree. He has published more than fifteen research articles in different international journals/conferences of repute. His research areas include reversible data hiding, cryptography and data mining.

 **Prof. Dr. Anirban Das** is associated with the University of Engineering & Management, Kolkata, as a Full Professor in Computer Science & Engineering. Prior to that, he was the founder HOD, CSE in Amity University Kolkata. Dr. Das started his career as a software engineer at HCL Technologies Ltd in 2007. He has completed his Doctoral degree in ICT from NIT Durgapur and is pursuing a Post Doctoral fellowship at Thu Dau Mot University, Vietnam. Dr. Das is a Visiting Scientist of University of Malaya, Malaysia. He is the honorary Vice President of the Scientific and Technical Research Association, Eurasia Research, USA; Fellow of Royal Society, UK; Fellow of Nikhil Bharat Siksha Parishad; Fellow of IETE, India; Fellow of ISRD UK; and Fellow of Indian Cloud Computing Association, India. He has authored 11 books; 1 research project funded by DSP-SERB, Govt. of India; 51 patents (Indian & International) filed; and more than 56 research publications, mostly in journals and conferences of international repute. He has served as Keynote Speaker/Invited speaker/ Session Chair in more than 30 conferences of international repute. He is the invitee as delegate from academics of several national bodies, such as CII, NASSCOM, FICCI and EDCN. He is nominated in WALL OF FAME of myGov as "THE CONFEDERATION OF ELITE ACADEMICIANS OF IICDC" powered by DST, Govt of India, AICTE and myGov, 2019. He is the 'Innovation Ambassador" certified by MHRD Innovation Cell, Govt. of India. He has a blended experience of almost 15 years in academics, research, and industry. He was awarded with Moulana Abul Kalam Azad Excellence Award of Education 2020; Inclusive Policy Lab Expert, UNESCO, USA, 2019; and University Best Teacher Award (UEM Kolkata), 2018. He is the recipient of Swami Vivekananda Yuba Samman 2015 by HCF, West Bengal, and Silver Champ Xtra-Miles Award 2009 by Employee First Council- HCL Technologies Ltd. His research interests include Blockchain Technologies, IoT, Business Forecasting and Machine Learning.

Contributors

Arijit Banerjee
Department of Information Technology
B.P. Poddar Institute of Management and
    Technology
Kolkata, India

Aritra Bhattacharjee
Department of Information Technology
B.P. Poddar Institute of Management and
    Technology
Kolkata, India

Akashdeep Bhattacharya
Department of Information Technology
B.P. Poddar Institute of Management and
    Technology
Kolkata, India

Semanti Chakraborty
Department of Electronics and
    Communication
Amity School of Engineering and
    Technology Kolkata
Amity University Kolkata
Kolkata, India

Shouvik Chakraborty
Department of Computer Science and
    Engineering
University of Kalyani, Kalyani
Nadia, West Bengal, India

Sankhadeep Chatterjee
Department of Computer Science &
    Engineering
University of Engineering & Management
Kolkata, India

Sudip Chatterjee
Department of Computer Science and
    Engineering
Amity School of Engineering and
    Technology Kolkata
Amity University Kolkata
Kolkata, India

Shyamapriya Chowdhury
Department of Information Technology
Narula Institute of Technology
Kolkata, India

Anupam Das
Department of Computer Science and
    Engineering
Amity School of Engineering and
    Technology Kolkata
Amity University Kolkata
Kolkata, India

Sachikanta Dash
Department of Computer Science and
    Engineering
DRIEMS Autonomous Engineering College
Cuttack, India

Stobak Dutta
Department of Computer Science and
    Engineering
University of Engineering and Management
Kolkata, India

Souvik Ghata
Department of Computer Science
University of Engineering and Management
Kolkata, India

Aninda Ghosh
Research and Development Cell
Altor
Bangalore, India

Medha Gupta
Amity Institute of Information
    Technology Kolkata
Amity University Kolkata
Kolkata, India

Shruti Harrison
Department of Computer Science
University of Engineering and
    Management
Kolkata, India

Gitosree Khan
Department of Information Technology
B.P. Poddar Institute of Management and
    Technology
Kolkata, India

D. Loganathan
Department of Computer Science and
    Engineering
Pondicherry Engineering College
Puducherry, India

Chittabrata Mal
Amity Institute of Biotechnology
Amity University Kolkata
Kolkata, India

Kalyani Mali
Department of Computer Science and
    Engineering
University of Kalyani, Kalyani
Nadia, West Bengal, India

Dhritiman Mukherjee
Amity School of Engineering and
    Technology Kolkata
Amity University Kolkata
Kolkata, India

Krishnendu Nandi
Department of Information Technology
B.P. Poddar Institute of Management and
    Technology
Kolkata, India

Neelamadhab Padhy
Department of Computer Science and
    Engineering
GIET University, Gunupur
Odisha, India

Kanik Palodhi
Department of Applied Optics and
    Photonics
University of Calcutta
Kolkata, India

Abhijit Paul
Amity Institute of Information
    Technology Kolkata
Amity University Kolkata
Kolkata, India

Subrata Paul
Department of Computer Science and
    Engineering MAKAUT, Kalyani
West Bengal, India

Tuhin Utsab Paul
IT/Computer Application
St. Xavier's University
Kolkata, India

Kinjal Raykarmakar
Department of Computer Science
University of Engineering and
    Management
Kolkata, India

Anil Saroliya
Department of Information Technology &
    Engineering
Amity University
Tashkent, Uzbekistan

Santosh Kumar Satapathy
Department of Computer Science and
    Engineering
Pondicherry Engineering College
Puducherry, India

Soumya Sen
A. K. Choudhury School of
    Information Technology
University of Calcutta
Kolkata, India

Kanishka Sharma
Department of Computer Science and
    Engineering
Mody University of Science and
    Technology
Rajasthan, India

Ananya Singh
School of Engineering and Technology
Mody University of Science and
  Technology
Rajasthan, India

Gour Sundar Mitra Thakur
Department of Computer Science and
  Engineering
Dr. B.C. Roy Engineering College
Durgapur, India

1

IoT in the Healthcare Industry

Semanti Chakraborty[1] **and Kanik Palodhi**[2]
[1]*Department of Electronics and Communication,*
Amity University, Kolkata, India
[2]*Department of Applied Optics and Photonics,*
University of Calcutta, India

CONTENTS

1.1 Introduction

The modern era is defined for its automation and industrialisation, with rapid exchange of human labour for its automation counterpart. This automation is driven by the presence of innumerable sensors and transducers across all walks of life. All these sensors generate electrical signals based on certain environmental conditions. For each condition, defined by parameters such as temperature or humidity, an electrical response is noted, giving rise to multiple datasets. According to the US National Science Foundation, approximately 20 billion sensors are going to be connected to the global network by 2020, which is an exponential growth to say the least [1–3].

Most companies now have a digital footprint, and as a consequence, huge numbers of objects, connected to various types of networks, will need efficient and precise data flow. Otherwise, the entire economy can be hampered due to a sudden glitch, since more and

DOI: 10.1201/9780429318078-1

more businesses are dependent on information coming from continuous monitoring of these connected objects. The basic concept is that humans are in touch with the world through connected objects distributed over a network. In other words, humans and their surroundings are connected through distributed sensors over a network of networks, the internet [4]. This concept of all-pervading connectivity was first proposed by British innovator Kevin Ashton in 1999. He coined the phrase "Internet of Things", or IoT, as it is often referred [5]. Though it was futuristic at that time, within 20 years IoT was standardised as a protocol in many applications [6–10].

After its inception and initial hiccups, IoT has slowly become a buzzword. The term now effectively amalgamates data acquisition, data storage and data analysis. Every day new applications are being designed and implemented since IoT has cut down the time to market for a product due to familiarity and format. Typical applications of IoT include the following, and this is in no way an exhaustive list:

1. Telecom [11–13]
2. ERP [14,15]
3. E-governance and law enforcement [16,17]
4. Smart home and smart city [18–20]
5. Renewable energy [21,22]
6. Healthcare and medical technologies [23–25]

This book chapter concentrates on the final one, covering the health sector. This sector more than any other domain has a telling effect on human lives, and it does so instantaneously. The domain is termed MIoT, or IoT applied in the field of medical technologies and healthcare, and many of the advanced countries are investing heavily to harness its benefits. A few of the plans are mentioned below:

1. China's National IoT Plan by the Ministry of Industry and IT
2. European Research Cluster on IoT (IERC)
3. Japan's u-Strategy
4. UK's Future Internet Initiatives
5. Italian National Project of Netergit
6. NIST report for cyber-physical systems and IoT in the United States [23,26–29]

The Indian government's (GoI) Digital India Mission includes a health portal and central storage of health-related medical data as well as image storage [30,31].

The predominant reason for this huge investment, particularly in the case of India, is the scarcity of suitable medical care and treatment for ordinary people, which will be discussed in Chapter 2 [32]. In recent times, private medicare has compounded the cost with an increase in insurance premiums.

This situation clearly points to the need for a drastic reduction in the number of patients who need to be seen in hospitals or clinics versus those who can use continuous health monitoring. In addition, it points to the importance of preventive medicine [33]. It calls for an emergency technology intervention, or "disruptive technology", and we believe MIoT is going to occupy that place. Previously, similar needs were catered to by using an ensemble of technologies that were referred to as "biotelemetry", or biomedical telemetry

[34,35]. Now, with the help of cloud computing, IoT has enabled the basic information to be uploaded on a platform that is interoperable and immediately accessible [32,36–39]. Apart from the reduction of patients who need to be in hospitals, other pertinent needs can be smoothly managed using this technology. In the following section, we discuss this technology in detail as well as a few future and special needs. This goes to show that in the presence of a simple, stable and predictable platform, many innovations can take place, driving the economy.

1.2 Need Analysis and Beneficiaries

In this section, we illustrate a need analysis with a few specific cases. Many sections of society are influenced by MIoT, and we should be prepared for a behavioural shift in medical management if this technology is introduced internationally [40–42].

Consider the relationship between patients and doctors. Globally, there is a lack of affordable healthcare for moderate income groups and the poor. Already, there is a huge emphasis on preventive medicine, which aims to monitor health or promote regular health check-ups. According to the Indian National Health Profile, a wing of Ministry of Health, GoI, only around 1 allopathic doctor per 11,000 patients is available. (This information was obtained from 2018 data.) This is far below the World Health Organization's global standard doctor and patient ratio of 1:1,000 [30,32]. There are many developed countries, even the United States, who do not have the best track record [32]. To improve the reach of the medical system and infrastructure to an individual level, connectivity and availability of appropriate sensors, or in short MIoT systems, need to expand.

The most important contributions, both in terms of finance and technology, should come from the government and the medical industry. According to a report by the International Data Corporation and US-based tech company Aeris, the investment in IoT is expected to be $1.2 trillion by the end of 2022 [43–45]. According to another report by Grand View Research, IoT spending in healthcare was $534 billion up to 2019; yet, in many cases, there is a need for immediate spending, particularly in rural healthcare and primary healthcare [46]. Switching to electronic health records and electronic medical records should save tremendous costs and provide more transparency. Hospital management and present healthcare personnel need to be sensitized to MIoT and the immediate improvement it can bring in [29,47–49].

According to John M. Halamka, MD, CIO of BIMDC, Boston, a respected medical analyst and blogger, MIoT requires 80 MB of data storage per patient per year, and this includes both text and images. Obviously, image data storage (generated from radiography, USG and CT scan, etc.) requirements are much more compared to text (structured and unstructured), and the storage is always a mix of different platforms [50]. Considering all of these factors, he mentioned in his blog that the typical cost (in 2011) was $1.27 per GB. It has now been reduced from that time due to multiple companies offering attractive features such as built-in statistical analysis and data security.

Finally, coming back to the need classification, a few are mentioned below. This list is in no way exhaustive, but it is based on present-day usage, which is changing at an exponential rate.

1. Monitoring in-house (i.e., within hospital from central facilities) or within home, particularly for older adults [51–53]
2. Data management of patients for administrative purposes [29,54,55]
3. Need for immediate delivery of data, results, or recommendations for faster decision making [23,56–58]
4. Special cases such as individuals with disabilities, animals, athletes and astronauts [59–64]

Looking at the above needs, it is clear that there are many stakeholders who should view MIoT as a necessary subject to research. Persons involved at different levels of healthcare industry starting from caregivers to policy heads should be ready to honestly analyse the implications of this field. As citizens whose lives will be affected by each move the healthcare industry makes, we should also be aware of these subtleties [27,29,65].

1.2.1 Monitoring In-House (i.e., within Hospital or Clinics)

Patients are constantly monitored using multiple sensors, such as thermometers, pulse oximeters, or electrochemical sensors. These sensors update current vital health parameters and help to predict the condition of patients within the hospital or clinic. Table 1.1 lists a few of these sensors and illustrates their respective measured parameters [65].

Apart from these, there are other electrochemical sensors for glucose measurements in blood, blood pressure measurement using capacitive or piezoelectric sensors, and so forth. These sensors in an IoT set-up are going to be placed in a wireless network, either local or distributed. These networks are referred to as WPAN (i.e., wireless personal area network) or WBAN (i.e., wireless body area network). Many of them are not standardised and act as ad-hoc networks. Therefore, in many important applications, these sensors are placed over standard WLAN and operated based on cloud or grid computing. This system is discussed in the final section.

1.2.2 Patients' Movement within Hospitals and Other Management

Another important need is tracking movement of patients and their associate devices/ documents as they move among different departments within the same premises or authority. This generates the need for a point-to-point tracking system with the help of radio frequency identification (RFID). This system is used for low-cost and efficient tracking and is commonly referred to as a "tag". It does not need line-of-sight communication and can be attached, or "tagged", to an object such as medical instruments or persons for examining their movement or status [23,65].

TABLE 1.1

Common Sensors Used for Vital Parameter Measurements

Parameters	Sensors
Heart rate	Electrodes, pulse oximeter
Body temperature	Digital thermometers, thermistors
Respiration rate	Strain gauges, piezo-resistive sensors
Saturation oxygen (SpO$_2$)	Pulse oximeter

These tags can be used to monitor continuous movement within a short range depending upon the antenna and frequency of RF wave used. For MIoT applications, a typical range is 100 ft [23,66,67]. This ensures the safety of roaming patients as well as associates in unauthorised areas. It is also important for security of the premise and its instruments. Many medical units now even issue tags to the visitors.

1.3 Doctors' Perspective

Doctors, the most important participants in the healthcare industry, also benefit from the use of MIoT. Doctors can instantly investigate variations in health parameter data compared to their normal ranges and even set specific alarms, which currently are performed manually. This continuous or 24×7 connectivity to health parameters of patients enables smooth decision making due to early diagnosis [68].

This will reduce the number of visits doctors need to make and save time. Particularly in a country such as India, where the doctor-to-patient ratio is abysmally low, this time can be used effectively.

1.4 Health Cost Reduction

In many cases, continuous parameter monitoring of sick patients as well as healthy persons, particularly athletes, is needed. These services were previously only available in labs. Due to a reduction in the size of sensors as well as the development of remote data transmission and analysis, health costs have been significantly reduced. This has lead to a surge in preventive medicine, where regular measurement of vital parameters could decrease the chances of chronic diseases [45,50,69,70].

1.5 Data Management of Patients for Administrative Purposes

Data management is strictly a part of information and communication technology; yet, IoT services have recently made in-roads into this domain due to cloud computing. Increasingly, data need to be omnipresent, particularly during emergencies [29,54,55]. In the case of medical units, apart from employee and internal matters, other data need to be stored and checked, These include identification data (for sick or injured persons and experts) and corresponding lab test reports with the latest vital parameter updates. These reports contain both imaging and non-imaging or clinical reports. In different parts of India, however, different practices and formats are followed, even for similar sets of tests or studies. This poses a huge challenge not only to medical practitioners but also to the general public [71,72].

The data collected from different hospitals and medical units are used for community-level statistical analysis. This analysis is very important, since many seasonal diseases can spread fast, and an endemic/epidemic situation may arise [73]. Apart from this, health budget projections and policies are formed based on this data analysis. A significant bioinformatics set-up is required for accurate predictions.

Insurance claims also require specific documentation and details. Certain expenses need to be formatted in a particular way, with a process flow where each step needs confirmation. These steps require access to a database that is clear, precise and unambiguous. A few examples of these processes are (a) tagging supporting evidence for each payment, (b) creating a liaison with insurance and data sharing, and (c) maximising profit by reducing costs through organisation of regular check-ups and monitoring. This also lowers the probability of terminal diseases, which generate significant costs [74,75].

1.6 IoT in the Healthcare Industry

Typically, sensors are connected to a wireless sensor network (WSN), with several nodes containing each sensor with their specific identifiers so that information can be processed and commands can be issued based on the situation, if required. In many cases, these commands are issued automatically, and there is little human intervention. The entire process is controlled from a centralised server within the hospital premises or from a remote server (i.e., using cloud services) [23,65,66].

The most common protocols in WSNs used for in-house monitoring are ZigBee and 6LoWPAN following IEEE 802.15.4 standard. Among these, 6LoWPAN, a short form of low-power wireless personal area network transmission following IPv6 protocol, is one of the most important systems in IoT. It is yet to be fully integrated to the mobile network; therefore, many WLAN (IEEE 802.11 onwards) networks are used for interoperability usage. There are new findings on integration of 6LoWPAN with MIPv6, which supports mobile connection; however, they have yet to become popular. The number of sensors attached to a WLAN depends on the data rate of the connection, available bandwidth, installation protocol and even the characteristics of the network interface. A comparison between them is provided below due to their varying speed and range [23,55,65,76–81] (Table 1.2).

As described earlier, RFID technology is also used for security access and other purposes in different medical units. Essentially, an RFID, or a tag as it is commonly referred, is a microchip that generates an RF signal through an antenna and a local server, then simultaneously verifies and tracks the signal to monitor the movement. RFIDs can be seamlessly integrated into WSN systems, too, so that robust and efficient systems can be designed [55].

TABLE 1.2

Different WLAN Technologies with Their Vital Parameters

Technology Standard (IEEE)	Band (GHz)	Typical Bit Rate (kbps)	Range (m)
802.11n	2.4	600	30
802.11ac	5.4	1,300	30
802.11ad	60	7,000,000	10
802.11ah	0.9	100	1,000

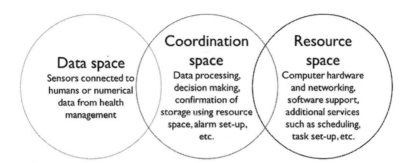

FIGURE 1.1
MIoT healthcare network set-up based on tasks.

IoT networks used in the healthcare industry can be separated into BAN/PAN and cloud or grid-based networks. In recent times, the trend is to be wireless; however, there are different connectivity levels across the domain. As shown in the diagram below (Figure 1.1), a typical network set-up can be sub-divided into three spaces based on the set of tasks performed [65].

Clearly, accurate and secure data delivery is essential for faster coordination to make critical and rapid decisions. For these requirements, three aspects of networks are of utmost importance, given below:

1. Topology
2. Architecture
3. Platforms

1.6.1 Topology

Network topology describes application-specific set-up of multiple sensors within a network for fast and secure data dissemination. According to Figure 1.1, this falls between data space and coordination space. Within this topology, multiple resources such as computers, laptops, and smartphones are turned into elements of networks (resources space). There can be multiple ways to connect; however, it is up to the network designer to effectively use the resources since there are different sensor signals such as ECG, BP, etc., coming from different parts of the body, and they may not always be compatible with network data protocol.

Apart from these networks, application of *grid computing* and *cloud computing* in MIoT is going to take the field to new heights. This is also under the domain of latest technology research. For all the above networks, incoming data is fed into *big data analytics* for trend analysis and decision making [29,82]. The topologies can be classified in terms of mobility, too, where data exchanges take place inside or outside a demarcated area. As discussed, sensors placed on the body or inside the premises are connected through wireless or wired connectivity.

There are, however, situations where communication with moving devices, such as devices on ambulances, wearable devices or medicine-delivery mechanisms, needs to be considered [65]. In these cases, GSM or WiMAX systems need to be embedded as well. This mobility distributed over the internet is demonstrated in a simple diagram, shown below [80,81] (Figure 1.2).

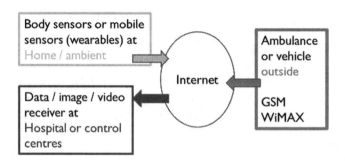

FIGURE 1.2
Heterogeneous network/HetNet.

This essentially is a heterogeneous network (HetNet), where data coming from different sensors are exchanged over the internet, typically using IPv6 [79,83,84]. There is a huge challenge in terms of maintaining interoperability and data security due to so much variety within one network [85,86]. One of the most important things about this entire topological application is to follow medical guidelines throughout each interaction with humans, whether they are health professionals or patients. They have to feel at ease with the use of technology and must be able to seamlessly interact with it.

1.6.2 Architecture

The architecture of the network provides a formal structure to the data coming from sensors to the final application support distributed over the network layers. A few technologies that are important and popular in MIoT considering the different ranges are the following:

1. WSN – wireless sensor network [81,87–90]
    i. Body area network [80]
    ii. Personal area network [91]
    iii. 6LoWPAN [77]
    iv. Bluetooth [92]
    v. Zigbee [93]
    vi. WiFi [94]

2. IPv6/MIPv6 [79]

An illustrative diagram is presented below following Gardas˘evic´ *et al.* [55]. Clearly, different communication technologies are adapted at different levels. Wireless technologies are the most popular MIoT technologies.

The network architecture will be an amalgamation of existing technologies such as TCP/IP, HTTP and WiFi with comparatively new technologies such as 6LoWPAN, Bluetooth or Zigbee. Designers mostly use well-known technologies such as HTTP or android at the application layer to create a seamless user experience (Figure 1.3).

e-Health delivery systems for community medical services have been proposed based on these architectures. Considering multiple stages of data transmission and

FIGURE 1.3
Intelligent network architecture.

interoperability, this intelligent network process is demonstrated in the figure above [95–97]. There are significant challenges ahead, predominantly based on the need for faster and better service.

For MIoT, IPv6-based 6LoWPAN is slowly making an indelible mark and becoming the most popular choice. It, however, does not support mobile IPv6 (MIPv6), which remains a challenge, and continuous research is happening in this field [23,65,97–99]. Even in this case, the front-end application supports are typically provided by the existing technologies, as already mentioned. An architecture known as V2IIn can be used as a dynamic network for moving vehicles [79]. This architecture typically follows a lightweight IPv6 protocol based on the sender–receiver model by following a loop.

REQUEST ↔ ACCEPT/RENEGOTIATE ↔ ACKNOWLEDGE/REFUSE ↔ READY

Here, there should be a discussion on middleware semantics, which are the processes that provide technical infrastructure for data transmission between two or more systems [55,86,97]. Typically they are categorised as

1. Application-specific
2. Agent-based
3. Virtual machine/VM-based
4. Tuple-spaces
5. Database-oriented
6. Service-oriented architecture (SOA)
7. Message-oriented middleware (MOM)

Among them, the first five have been deployed for specific applications, and they have issues in scalability and manageability over a distributed, larger network. Healthcare industries, particularly within medical units and ambient assisted living (AAL) under WBAN/WPAN, have adopted SOA for most of the applications [100]. Two of the most important SOAs are the SPHERE project [101] and REACTION [102], which has been successfully used with medical applications.

Similar projects are currently under research for smart home applications. In this case, the data providers and data consumers are bound by a contract; therefore, information flow depends on communication systems components. This is where a completely information-based, message-oriented system has an edge provided by MOM architecture over a distributed network. It is compatible with different network protocols and fully scalable under a publish-subscribe operation scheme. This architecture is, therefore, emerging as the middleware semantic for MIoT. The most common data models used in the middleware are written below [23,86,101,103]:

TABLE 1.3

Middleware Projects in MIoT and AAL

Name of the Project	Architecture	Semantic	Application
Linksmart	SOA	Sensors	MIoT – diabetes monitoring
SM4ALL	SOA	Sensors	AAL for older adults or individuals with disabilities
Mango	SOA	Sensors	MIoT – diabetes monitoring
xAAL	MOM	Sensors and messages	AAL and home automation

    i. XML

    ii. Web Service Definition Language (WSDL)

    iii. JavaScript Object Notation (JSON)

    iv. Resource Description Framework (RDF)and

    v. Web Ontology Language (OWL)

A few middleware projects are described in Table 1.3; they are used in MIoT or AAL and have gained significant popularity [86].

Despite these methods, there are huge challenges to the infrastructure:

1. Interoperability among heterogeneous devices

2. Scalability for many connected devices

3. Automated interactions of objects and devices

4. Heterogeneity in network infrastructure (wired and wireless, etc.)

5. Abstraction

Many modern systems are trying to overcome these challenges, and we hope a much better heterogeneous architecture suitable for MIoT will emerge soon. Many of these challenges are not only architectural but also need to be addressed by networks and platforms, described in the next section.

1.7 Platforms

MIoT platforms are computational software or networks that provide an interface to users for requesting raw data or processed data and/or services [23,55,86,104]. These requests include:

1. Real-time data coming from sensors or nodes for monitoring and control of MIoT/AAL applications

2. Real-time connectivity data for smooth running of medical units

3. Stored data for EHRs (e-health records)

4. Instrument health monitoring data

5. Security checks

6. Statistical pattern or trend analysis

More and more, these data are fetched over wireless network or HetNet, and the demand for data is at an all-time high. Understandably, user requests across different platforms will need immediate standardisation and interoperability. A resource control mechanism should also be built into the system for optimisation of data sharing and security measures. Considering these constraints, there are two categories of operation, given in Table 1.4 [86].

Multiple papers focus on EHRs that describe residents' healthcare and distributed hierarchical platforms based on smart objects. A few techniques also focus on interoperable systems across HetNet and provide greater user friendliness. One of the techniques presents automatic design methodology (ADM) along with many of the challenges, such as optimisation and platform management [105].

1.7.1 Applications and Services

In MIoT, there is a thin line between classification of applications and services. Essentially, applications are closer to data acquisition and preliminary levels of processing, whereas services are combinations of multiple applications arranged to complete a bigger task. Here, bigger tasks involve multilevel networks and multi-parameter analysis. Considering the number of users, applications can be classified into two groups, as shown in the following points [23,54,65]:

   a. *Sensor level,* where a sensor set consisting of only a few sensors is used, mostly for personal applications
   b. *Community level,* where many sets of sensors are used to serve a bigger community

In the following table, a possible distinction between applications and services is made. A set of MIoT applications is separated, and corresponding uses are clubbed together as services (Table 1.5).

As can be seen in the above table, personal smartphone assistance services, or smartphone apps that provide day-to-day health updates using data coming from wearable sensors connected to smart watches or shoes, use few sensor-level applications such as temperature sensors or pulse monitors. Simple computational tasks, such as thresholding for alarms or notifications, can also be set since smartphones have quite a bit of processing capability.

TABLE 1.4

Categories of Operation Related to the Application of Platforms

Service Providers	Users
1. Healthcare (medical units)	1. Patients
2. Application designers	2. Data storage (repository in medical units, etc.)
3. Connectivity	3. Doctors or auditors, etc.
4. Platform	
5. Content, etc.	

TABLE 1.5

Features of MIoT Applications and Services

MIoT Applications		Services – Combines Multiple Applications
Sensor Level	**Community Level**	
1. Blood pressure	1. Patients' monitoring	1. Personal smartphone assistance
2. Glucose monitoring	2. Older persons' and	2. Medical unit/hospital
3. Temperature	children's	management systems
monitoring	monitoring	3. Ambient assisted living (AAL)
4. ECG	3. Drug/chronic	4. Community-level medical
5. Pulse oximetry	disease treatment	information processing systems for
	4. Emergency	governments and predictions/
		projections for budgeting
		5. Completely automated digital
		assistant or advisors

TABLE 1.6

MIoT Services

Name of the App	Purpose	Application	Sensors
Prognosis	Diagnosis	Y	N
Johns Hopkins antibiotic guide	Diagnosis	Y	N
FDA drugs	Drug reference	Y	N
Epocrates	Diagnosis, medical calculation	Y	N
Voalté One	Communication	Y	N
Google Fit	General health monitoring	Y	Y
iOximeter	General health monitoring	Y	Y

The same data collected from multiple sources can be used at the community level for medical information processing systems. An example is multi-parameter thresholding for monitoring critically of patients and informing health experts or doctors. Automated fetching of the medical records of these patients is even possible. At the government or policy-making level, a data repository for research on bioinformatics for trend analysis has been planned and implemented. In Table 1.6, a few applications and services commonly available on smartphones are provided to illustrate this point [23,55].

These applications are clearly data intensive and do not require real-time sensor data; however, personal health monitoring embedded sensors are accessed by the applications. More and more data analytics are getting built into these systems, enabling faster decision making.

1.8 Conclusion

In this chapter, we have reviewed the current state of IoT applied to the field of healthcare and medical technologies. This field has seen an exponential growth in terms of products and research at different levels, such as sensors, networks and repositories. It is astounding to note that within the last few years billions of dollars have been poured into the field, both from governments and from private entities, for developing capabilities for welfare and business, respectively. Being an interdisciplinary field, these capabilities retain applicability in multiple domains. The emergence of wearables and their integration to smart devices have ignited a new generation of product ideas and will continue to drive R&D in this direction.

One of the main drivers of this technology is the abundance of smartphone users present among different strata within society who use touchscreen apps to demand and use these services instantly. Availability of embedded sensors at low costs has also driven the demand for real-time sensor data. The ability of output from these sensors to feed into the nodes of networks has also sparked the possibility of real-time heterogeneous architecture in MIoT.

There are, however, a lot of challenges ahead to create a robust and accurate IoT framework. Standardisation of the heterogeneity and interoperability among multiple systems remains a fundamental problem. Characterising the static and dynamic nature of sensor nodes and choosing an appropriate data analytic model for different systems are new fields of research. Internet and communication technology is also another associated field that is offering many options. Therefore, this is an exciting time to be a researcher or an engineer in this field.

References

1. "Sensor-Based Economy" *Wired*. https://www.wired.com/brandlab/2017/01/sensor-based-economy/
2. Analysys Mason, "Imagine an M2M World with 2.1 Billion Connected Things", 2011. http://www.analysysmason.com/about-us/news/insight/M2M_forecast_Jan2011/
3. Q. F. Hassan, "Internet of Things in Smart Ambulance and Emergency Medicine", in *Internet of Things A to Z: Technologies and Applications*, IEEE, pp. 475–506, 2018.
4. ITU-T Internet Reports, "Internet of Things", (Global Report), 2005.
5. K. Ashton, "That 'Internet of Things' Thing: In the Real World Things Matter More than Ideas", *RFID Journal*(http://www.rfidjournal.com/articles/view?4986), 2009.
6. G. Kortuem, F. Kawsar, D. Fitton, and V. Sundramoorthy, "Smart Objects as Building Blocks for the Internet of Things", *IEEE Internet Computing*, vol. 14, no. 1, pp. 4451, 2010.
7. M. Weiser, "The Computer for the 21st Century", *Scientific American*, Sept., 1991, pp. 94–104; reprinted in IEEE Pervasive Computing, Jan.–Mar. 2002, pp. 19–25.
8. World Economic Forum, "The Global Information Technology Report 2012—Living in a Hyperconnected World", 2012. http://www3.weforum.org/docs/Global_IT_Report_2012.pdf
9. J. Höller, V. Tsiatsis, C. Mulligan, S. Karnouskos, S. Avesand, and D. Boyle, "From Machine-to-Machine to the Internet of Things: Introduction to a New Age of Intelligence", Amsterdam, The Netherlands: Elsevier, 2014.
10. W. Arden, M. Brillouët, P. Cogez, M. Graef, B. Huizing and R. Mahnkopf, "More than Moore, White Paper", 2010. http://www.itrs2.net/uploads/4/9/7/7/49775221/irc-itrs-mtm-v2_3.pdf
11. I. Khan, F. Belqasmi, R. Glitho, N. Crespi, M. Morrow, and P. Polakos, "Wireless Sensor Network Virtualization: A Survey", *IEEE Communications Surveys & Tutorials*, vol. 18, no. 1, pp. 553–576, 2016.

12. International Telecommunication Union—ITU-T Y.2060—(06/2012)—Next Generation Networks—Frameworks and functional architecture models—Overview of the Internet of things.

13. J. T. J. Penttinen, "Internet of Things", in *Wireless Communications Security: Solutions for the Internet of Things*, Wiley, pp.112–149, 2015.

14. M. Kärkkäinen and J. Holmström, "Wireless Product Identification: Enabler for Handling Efficiency, Customisation and Information Sharing", *Supply Chain Management: An International Journal*, vol. 7, pp. 242–252, 2002.

15. H. Geng, "Internet of Things Business Model", In *Internet of Things and Data Analytics Handbook*, Wiley Publication, pp. 735–757, 2017.

16. S. Chen, C. Yang, J. Li, and F. R. Yu, "Full Lifecycle Infrastructure Management System for Smart Cities: A Narrow Band IoT-Based Platform", *IEEE Internet of Things Journal*, vol. 6, no. 5, pp. 8818–8825, 2019.

17. L. Caviglione, S. Wendzel, and W. Mazurczyk, "The Future of Digital Forensics: Challenges and the Road Ahead", *IEEE Security & Privacy*, vol. 15, no. 6, pp. 12–17, 2017.

18. Frost & Sullivan, "Mega Trends: Smart is the New Green" Blog, 2019. www.frost.com.

19. OUTSMART, FP7 EU project, part of the Future Internet Private Public Partnership, "OUTSMART—Provisioning of Urban/Regional Smart Services and Business Models Enabled by the Future Internet", 2018. http://www.fi-ppp-outsmart.eu/enuk/Pages/default.aspx 26-08-2018.

20. L. Sánchez, J. Antonio, G. Veronica, T. Garcia, et al. "SmartSantander: The Meeting Point Between Future Internet Research and Experimentation and the Smart Cities", *Project: SmartSantander*, FNMS Summet, 2011.

21. Q. F. Hassan, "Implementing the Internet of Things for Renewable Energy", in *Internet of Things A to Z: Technologies and Applications*, IEEE, pp. 425–446, 2018.

22. O. Vermesan, et al., "Internet of Energy—Connecting Energy Anywhere Anytime", In *Advanced Microsystems for Automotive Applications 2011: Smart Systems for Electric, Safe and Networked Mobility*, Springer, Berlin, 2011, ISBN 978-36-42213-80-9.

23. S. M. Riazul Islam, K. Daehan, K. Humaun, M. Hossain, and K., Kyung-Sup, "The internet of things for health care: A comprehensive survey", IEEE Access, vol. 3, 678–708, 2015.

24. H. Fang, X. Dan, and S. Shaowu, "On the Application of the Internet of Things in the Field of Medical and Health Care", Green Computing and Communications (GreenCom), 2013 IEEE and Internet of Things (iThings/CPSCom), IEEE International Conference on and IEEE Cyber, Physical and Social Computing, pp. 2053–2058, 2013

25. D. Ugrenovic, and G. Gardasevic, "Performance Analysis of IoT Wireless Sensor Networks for Healthcare Applications", In The 2nd International Conference on Electrical, Electronic and Computing Engineering, IcETRAN 2015, Silver Lake, Serbia, 2015.

26. A. Gluhak, S. Krco, M. Nati, D. Pfisterer, N. Mitton, and T. Razafindralambo, "A Survey on Facilities for Experimental Internet of Things Research," *IEEE Communications Magazine*, vol. 49, no. 11, Nov. 2011, pp. 58–67.

27. K. Ullah, M. A. Shah, and S. Zhang, "Effective Ways to Use Internet of Things in the Field of Medical and Smart Health Care", 2016 International Conference on Intelligent Systems Engineering (ICISE), Islamabad, pp. 372–379, 2016.

28. Workshop Report, Foundations for Innovation in Cyber-Physical Systems, NIST, 2019. https://nvlpubs.nist.gov/nistpubs

29. D. V. Dimitrov, "Medical Internet of Things and Big Data in Healthcare", *Healthcare Informatics Research*, vol. 22, no. 3, pp. 156–163, 2016.

30. Report (from) Central Bureau of Health Intelligence, "National Health Profile (NHP) - 2018", 13th issue.

31. "Draft Policy on Internet of Things", Ministry of Electronics and Information Technology (MeiTY), Govt. of Indiahttps://meity.gov.in/2015.

32. "Density of Physicians", Global Health Observatory (GHO) Data. Situation and Trends, http://www.who.int/ 2018.

33. Report from International Data Corporation, "IDC Forecasts Worldwide Spending on the Internet of Things to Reach \$745 Billion in 2019, Led by the Manufacturing, Consumer, Transportation, and Utilities Sectors", 2019, www.idc.com/

34. K. S. Nikita, "Introduction to Biomedical Telemetry", In *Handbook of Biomedical Telemetry*, Ed. Konstantina S. Nikita, John Wiley & Sons, Inc., 2014.

35. K. S. Nikita, J. C. Lin, D. I. Fotiadis, and M. T. Arredondo Waldmeyer, "Special Issue on Mobile and Wireless Technologies for Healthcare Delivery', *IEEE Transactions on Biomedical Engineering*, vol. 59, no. 11, pp. 3083–3270, 2012.

36. K. Romer, B. Ostermaier, F. Mattern, M. Fahrmair, and W. Kellerer, "Real-Time Search for Real-World Entities: A Survey", *Proceedings of the IEEE*, vol. 98, no. 11, pp. 1887–1902, 2010.

37. D. Guinard, V. Trifa, and E. Wilde, "A Resource Oriented Architecture for the Web of Things", In *Proceedings of the Internet Things (IOT)*, Nov./Dec. 2010, pp. 1–8.

38. L. Tan and N. Wang, "Future Internet: The Internet of Things", In *Proceedings of the 3rd International Conference on Advance Computer Theory Engineering (ICACTE)*, August 2010.

39. Z. Pang, "Technologies and Architectures of the Internet-of-Things (IoT) for Health and Well-being", M. S. thesis, Department of Electrical Engineering and Computer Science. KTH Royal Institute of. Technology, Stockholm, Sweden, 2013.

40. K. Vasanth and J. Sbert, "Creating solutions for health through technology innovation", *Texas Instruments*online at http://www.ti.com/lit/wp/sszy006/sszy006.pdf

41. J. Ko, C. Lu, M. B. Srivastava, J. A. Stankovic, A. Terzis, and M. Welsh, "Wireless Sensor Networks for Healthcare", *Proceedings of the IEEE*, vol. 98, no. 11, pp. 1947–1960, 2010.

42. H. Alemdar and C. Ersoy, "Wireless Sensor Networks for Healthcare: A Survey", *Computer Network*, vol. 54, no. 15, pp. 2688–2710, 2010.

43. "Internet of Things Ecosystem and Trends", IDC report or Worldwide Semi-annual Internet of Things Spending Guide, 2020.

44. D. Bandyopadhyay and J. Sen, "Internet of Things: Applications and Challenges in Technology and Standardization", *Wireless Personal Communications*, vol. 58, pp. 49–69, 2011.

45. N. L. Fantana et al., "IoT Applications—Value Creation for Industry", In *Internet of Things: Converging Technologies for Smart Environments and Integrated Ecosystems*, Eds. Ovidiu Vermesan and Peter Friess, River Publishers, 2013.

46. "IoT in Healthcare Market Worth \$534.3 Billion By 2025 CAGR: 19.9%". https://www.grandviewresearch.com/press-release/global-iot-in-healthcare-market

47. C. Auffray, R. Balling, I. Barroso, L. Bencze, M. Benson, J. Bergeron, et al. "Making Sense of Big Data in Health Research: Towards an EU Action Plan", *Genome Med*, vol. 8, no. 1, p. 71, 2016.

48. B. L. Filkins, J. Y. Kim, B. Roberts, W. Armstrong, M. A. Miller, M. L. Hultner, et al. "Privacy and Security in the Era of Digital Health: What Should Translational Researchers Know and Do About It?", *American Journal of Translational Research*, vol. 8, no. 3, pp. 1560–1580, 2016.

49. W. O. Hackl, "Intelligent Re-Use of Nursing Routine Data: Opportunities and Challenges", *Studies in Health Technology and Informatics*, vol. 225, pp. 727–728, 2016.

50. Personal blog of Dr. Halamka. http://geekdoctor.blogspot.com/2011/04/cost-of-storing-patient-records.html

51. A. Holzinger, M. Ziefle, and C. Röcker "Human-Computer Interaction and Usability Engineering for Elderly (HCI4AGING): Introduction to the Special Thematic Session". In:Miesenberger, K., Klaus, J., Zagler, W., Karshmer, A. (eds.) ICCHP 2010. LNCS, vol. 6180, pp. 556–559. Springer, Heidelberg, 2010.

52. I. Iliev and I. Dotsinsky "Assisted living systems for elderly and disabled people", short review. *Bioautomation*, vol. 15/2, no. 1314-1902, pp. 131–139, 2011.

53. V. Emmanouela et al. "Maintaining Mental Wellbeing of Elderly at Home, from 'Enhanced Living Environments Algorithms, Architectures, Platforms, and Systems", Ivan Ganchev, Nuno M. Garcia, Ciprian Dobre and Constandinos X. Mavromoustakis Rossitza Goleva (Eds.). Springer, 2018.

54. O. Vermesan et al., "Internet of Things Strategic Research and Innovation Agenda" In *'Internet of Things: Converging Technologies for Smart Environments and Integrated Ecosystems,'* Dr. Ovidiu Vermesan and Dr. Peter Friess (Eds.). River Publishers, 2013.

55. G. Gardas̆evic̆ et al., "The IoT Architectural Framework, Design Issues and Application Domains", *Wireless Personal Communication*, vol. 92, pp. 127–148, 2017. Doi. 10.1007/s11277-016-3842-3.

56. M. Wiesner and D. Pfeifer, "Health recommender systems: Concepts, requirements, technical basics and challenges", *International Journal of Environmental Research Public Health*, vol. 11, no. 3, pp. 2580–2607, 2014.

57. V. Trajkovik, S. Koceski, E. Vlahu-Gjorgievska, and I. Kulev, "Evaluation of health care system model based on collaborative algorithms", In Adibi, S. (ed.) Mobile Health. SSB, vol. 5, pp. 429–451.

58. S. Cvetanka et al., Development and Evaluation of Methodology for Personal Recommendations Applicable in Connected Health, from 'Enhanced Living Environments Algorithms, Architectures, Platforms, and Systems,' Eds. Ivan Ganchev, Nuno M. Garcia Ciprian Dobre, Constandinos X. Mavromoustakis. Rossitza Goleva, Springer, 2018

59. C. Coelho, D. Coelho, and M. Wolf, "An IoT smart home architecture for long-term care of people with special needs", 2015 IEEE 2nd World Forum on Internet of Things (WF-IoT), Milan, 2015. pp. 626–627.

60. L. Nóbrega, A. Tavares, A. Cardoso, and P. Gonçalves, "Animal monitoring based on IoT technologies", 2018 IoT Vertical and Topical Summit on Agriculture - Tuscany (IOT Tuscany), Tuscany, 2018, pp. 1–5.

61. Y. Wang, M. Chen, X. Wang, R. H. M. Chan, and W. J. Li, "IoT for Next-Generation Racket Sports Training", *IEEE Internet of Things Journal*, vol. 5, no. 6, pp. 4558–4566, 2018.

62. S. Kubler, J. Robert, A. Hefnawy, K. Främling, C. Cherifi, and A. Bouras, "Open IoT Ecosystem for Sporting Event Management", *IEEE Access*, vol. 5, pp. 7064–7079, 2017.

63. K. Tseng, L. Liu, C. Wang, K. L. Yung, W. H. Ip, and C. Hsu, "Robust Multistage ECG Identification for Astronaut Spacesuits With IoT Applications," IEEE Access, vol. 7, pp. 111662–111677, 2019.

64. M. Canina, D. J. Newman, and G. L. Trotti, "Preliminary considerations for wearable sensors for astronauts in exploration scenarios," 2006 3rd IEEE/EMBS International Summer School on Medical Devices and Biosensors, Cambridge, MA, 2006, pp. 16–19.

65. V. A. George et al. "Healthcare Sensing and Monitoring", In *Enhanced Living Environments Algorithms, Architectures, Platforms, and Systems*, Eds. Ivan Ganchev et al., Springer Open, 2019.

66. A. Mitrokotsa and C. Douligeris, "Integrated RFID and sensor networks: Architectures and applications. In: RFID and Sensor Networks: Architectures, Protocols, Security, and Integrations", pp. 511–536. CRC Press, Taylor & Francis Group, Boca Raton, 2010.

67. X. Liu, J. Yin, Y. Liu, S. Zhang, S. Guo, and K. Wang, "Vital Signs Monitoring with RFID: Opportunities and Challenges," *IEEE Network*, vol. 33, no. 4, pp. 126–132, July/August 2019.

68. V. Emmanouela et al. "Maintaining Mental Wellbeing of Elderly at Home" in Enhanced Living Environments Algorithms, Architectures, Platforms, and Systems, Eds. Ivan Ganchev et al, Springer Open, 2019

69. D. Niewolny, "How the Internet of Things is revolutionizing healthcare", 2013, online at http://freescale.com/healthcare, 2013.

70. A. Galletta, L. Carnevale, A. Bramanti, and M. Fazio, "An Innovative Methodology for Big Data Visualization for Telemedicine," *IEEE Transactions on Industrial Informatics*, vol. 15, no. 1, pp. 490–497, 2019.

71. M. N. Alraja, M. M. J. Farooqueand, and B. Khashab, "The Effect of Security, Privacy, Familiarity, and Trust on Users' Attitudes Toward the Use of the IoT-Based Healthcare: The Mediation Role of Risk Perception," *IEEE Access*, vol. 7, pp. 111341–111354.

72. T. Laksanasopin, P. Suwansri, P. Juntaramaha, K. Keminganithi, and T. Achalakul, "Point-of-Care Device and Internet of Things Platform for Chronic Disease Management," 2018 2nd International Conference on Biomedical Engineering (IBIOMED), Kuta, 2018, pp. 80–83.

73. H. Sato, K. Ide and A. Namatame, "Agent-based infectious diffusion simulation using Japanese domestic human mobility data as metapopulation network", 2016 Second Asian Conference on Defence Technology (ACDT), Chiang Mai, 2016, pp. 93–97.

74. A. A. Vazirani, O. O'Donoghue, and D. Brindley, "Meinert E Implementing Blockchains for Efficient Health Care", Systematic Review, J Med Internet Res, vol. 21, no. 2, 2019, p. e12439

75. G. Shobana and M. Suguna, "Block Chain Technology towards Identity Management in Health Care Application," 2019 Third International conference on I-SMAC (IoT in Social, Mobile, Analytics and Cloud) (I-SMAC), Palladam, India, 2019, pp. 531–535.

76. B. Alessandro et al., "A Common Architectural Approach for IoT Empowerment" in 'Internet of Things: Converging Technologies for Smart Environments and Integrated Ecosystems,' Eds. Ovidiu Vermesan and Peter Friess, River Publishers, 2013

77. M. S. Shahamabadi, B. B. M. Ali, P. Varahram, and A. J. Jara, "A network mobility solution based on 6LoWPAN hospital wireless sensor network (NEMO-HWSN)", in Proc. 7th Int. Conf. Innov. Mobile Internet Services Ubiquitous Comput. (IMIS), 2013, pp. 433–438.

78. H. Viswanathan, E. K. Lee, and D. Pompili, "Mobile grid computing for data- and patient-centric ubiquitous healthcare", in Proc. 1st IEEE Workshop Enabling Technology Smartphone Internet Things (ETSIoT), 2012, pp. 36–41.

79. S. Imadali, A. Karanasiou, A. Petrescu, I. Sifniadis, V. Veque, and P. Angelidis, "eHealth service support in IPv6 vehicular networks", in Proc. IEEE Int. Conf. Wireless Mobile Comput., Network Communication (WiMob), 2012, pp. 579–585.

80. N. Bui, N. Bressan, and M. Zorzi, "Interconnection of body area networks to a communications infrastructure: An architectural study", in Proc. 18th Eur. Wireless Conf. Eur. Wireless, 2012, pp. 18–24.

81. C. Alcaraz, P. Najera, J. Lopez, and R. Roman, "Wireless sensor networks and the Internet of Things: Do we need a complete integration?", in Proc. 1st Int. Workshop Security Internet Things (SecIoT), 2010.

82. M. Diaz, G. Juan, O. Lucas, and A. Ryuga, "Big data on the Internet of Things: An example for the e-health", in Proc. Int. Conf. Innov. Mobile Internet Services Ubiquitous Computing (IMIS), 2012, pp. 898–900.

83. L. You, C. Liu, and S. Tong, "Community medical network (CMN): Architecture and implementation", in Proc. Global Mobile Congress (GMC), 2011, pp. 16–21.

84. P. Swiatek and A. Rucinski, "IoT as a service system for eHealth", in Proc. IEEE Int. Conf. eHealth Network, Appl. Services (Healthcom), 2013, pp. 81–84.

85. J.-M. Bohli et al. "Security and Privacy Challenge in Data Aggregation for the IoT in Smart Cities", In 'Internet of Things: Converging Technologies for Smart Environments and Integrated Ecosystems,' Eds. Dr. Ovidiu Vermesan and Dr. Peter Friess, River Publishers, 2013.

86. Z. Rita et al. "Semantic Middleware Architectures for IoT Healthcare Applications", In Enhanced Living Environments Algorithms, Architectures, Platforms, and Systems, Eds. Ivan Ganchev et al., Springer Open, 2019

87. I. F. Akyildiz, W. S. Y. Sankarasubramaniam, E. Cayirci, "A survey on sensor networks", IEEE Communications Magazine, vol. 40, pp. 102–114, 2002

88. M.Johnson, et al., "A Comparative Review of Wireless Sensor Network Mote Technologies", In 8th IEEE Conference on Sensors (IEEE SENSORS 2009), Christchurch, New Zealand, 2009.

89. S. Ramesh, "A Protocol Architecture for Wireless Sensor Networks", Research Paper, School of Computing, University of Utah, 2008.

90. M. Kocakulak and I. Butun, "Overview of Wireless Sensor Networks towards IoT", In: IEEE 7th Annual Computing and Communication Workshop and Conference (CCWC), 2017.

91. Y. Turk and A. Akman, "Management of Low Powered Personal Area Networks Using Compression in SNMPv3", 2018 IEEE/ACS 15th International Conference on Computer Systems and Applications (AICCSA), Aqaba, pp. 1–5, 2018.

92. P. P. Ray, N. Thapa, and D. Dash, "Implementation and Performance Analysis of Interoperable and Heterogeneous IoT-Edge Gateway for Pervasive Wellness Care", In *IEEE Transactions on Consumer Electronics*, vol. 65, no. 4, pp. 464–473, 2019.

93. E. Spanò, S. Di Pascoli, and G. Iannaccone, "Low-Power Wearable ECG Monitoring System for Multiple-Patient Remote Monitoring", *IEEE Sensors Journal*, vol. 16, no. 13, pp. 5452–5462, 2016.

94. V. Bianchi, M. Bassoli, G. Lombardo, P. Fornacciari, M. Mordonini, and I. De Munari, "IoT Wearable Sensor and Deep Learning: An Integrated Approach for Personalized Human Activity Recognition in a Smart Home Environment", *IEEE Internet of Things Journal*, vol. 6, no. 5, pp. 8553–8562, 2019.

95. F. Firouzi, B. Farahani, M. Ibrahim, and K. Chakrabarty, "Keynote Paper: From EDA to IoT eHealth: Promises, Challenges, and Solutions", *IEEE Transactions on Computer-Aided Design of Integrated Circuits and Systems*, vol. 37, no. 12, pp. 2965–2978, 2018.

96. F. Hudson and C. Clark, "Wearables and Medical Interoperability: The Evolving Frontier", *Computer*, vol. 51, no. 9, pp. 86–90, 2018.

97. V. H. Díaz et al., "Semantic as an Interoperability Enabler in Internet of Things" In *Internet of Things: Converging Technologies for Smart Environments and Integrated Ecosystems*, Eds. Dr. Ovidiu Vermesan and Dr. Peter Friess, River Publishers, 2013.

98. A. J. J. Valera, M. A. Zamora, and A. F. G. Skarmeta, "An Architecture Based on Internet of Things to Support Mobility and Security in Medical Environments", 2010 7th IEEE Consumer Communications and Networking Conference, Las Vegas, NV, 2010, pp. 1–5.

99. D. Minoli, "Building the Internet of Things with IPv6 and MIPv6: The Evolving World of M2M Communications" (1st. ed.), Wiley Publishing, USA, 2013.

100. H. Mshali, T. Lemlouma, M. Moloney, D. Magoni, "A Survey on Health Monitoring Systems for Health Smart Homes", HAL Id: hal-01715576 'https://hal.archives-ouvertes.fr/hal-01715576', 2018.

101. A. Elsts et al., "Enabling Healthcare in Smart Homes: The SPHERE IoT Network Infrastructure", *IEEE Communications Magazine*, vol. 56, no. 12, pp. 164–170, 2018.

102. M. Ahlsén, S. Asanin, P. Kool, P. Rosengren, and J. Thestrup, "Service-Oriented Middleware Architecture for Mobile Personal Health Monitoring", Eds. K. S. Nikita, J. C. Lin, D. I. Fotiadis, M. T. Arredondo Waldmeyer *Wireless Mobile Communication and Healthcare. MobiHealth 2011, Lecture Notes of the Institute for Computer Sciences, Social Informatics and Telecommunications Engineering*, vol 83. Springer, Berlin, Heidelberg.

103. K. Petteri et al., "RDF Stores for Enhanced Living Environments: An Overview" In *Enhanced Living Environments Algorithms, Architectures, Platforms, and Systems*, Eds. Ivan, Ganchev et al., Springer Open, 2019.

104. D.-O. Kadian et al., "Towards a Deeper Understanding of the Behavioural Implications of Bidirectional Activity-Based Ambient Displays in Ambient Assisted Living Environments", In *Enhanced Living Environments Algorithms, Architectures, Platforms, and Systems*, Eds. Ivan Ganchev et al., Springer Open, 2019.

105. Y. J. Fan, Y. H. Yin, L. D. Xu, Y. Zeng, and F. Wu, "IoT-based Smart Rehabilitation System", *IEEE Transactions on Industrial Informatics*, vol. 10, no. 2, pp. 1568–1577, 2014.

2

Asynchronous BCI (Brain–Computer Interface) Switch for an Alarm

**Gitosree Khan, Arijit Banerjee, Aritra Bhattacharjee,
Krishnendu Nandi and Akashdeep Bhattacharya**

*B.P. Poddar Institute of Management &
Technology, Kolkata, India*

CONTENTS

2.1 Introduction

BCI, or brain–computer interface, is a specialized device that captures signals emitted directly from the brain and uses these signal datasets to execute computerized actions. Patients who experience tetraplegia, amyotrophic lateral sclerosis (ALS), other severe physical disability or inability to speak will benefit from this technology. They will be able to communicate with an attendant through an alarm whenever they want as well as in emergency situations (panic, rage), when the switch will be triggered to indicate a state of distress. This design is relatively simple and portable, and it is cost effective in setup and usage. The brain signals will be detected in real time and will trigger an alarm

DOI: 10.1201/9780429318078-2

under suitable parameters. The idea is to read the brain signals with the help of an electroencephalogram (EEG) device placed on the scalp. The EEG device may have many electrodes (multi-channel) or a single active electrode on a specific part of the scalp (single channel). We primarily target signals coming from the motor cortex because these regions of the brain show the most significant and readable changes in brain signal patterns, ensuring higher chances of success with our detection model. A way to voluntarily trigger the switch would be to think of moving either of the arms, the legs or the tongue (the four most commonly used MI [motor imagery] classes). This will invoke the motor cortex of the brain, resulting in a spike of beta waves in the brain signals that will trigger the switch. As mentioned above, this model also features a function to trigger the switch if the patient suddenly faces a panic attack or any other sudden distress or high-energy reaction.

The model of BCI proposed in this chapter is based on EEG and MI. We place single or multiple electrode receivers on the head and have one electrode connected to the earlobe that acts as the reference electrode. We have another electrode placed behind the head that works as the ground electrode to complete the circuit. The single or multiple active electrodes pick up the brain signals from where they are placed. The ground electrode completes the circuit while the potential difference between the active and the reference electrode is outputted.

Receiving the brain signals effectively using our interface is crucial. Different noise might be caused by electrical circuits in our interface. We might also need calibration to selectively detect the brain waves to apply our classifier on. Generally, multi-channel BCI is used to fetch brain signals. Irrelevant channels or channels with high noise levels will be excluded by running EEG channel selection, providing us with a better model [1]. However, multi-channel systems need an extra step for selecting the most useful dataset from an array of channels, which increases the computational costs. This also increases the complexity of the switch design. There is ongoing research on how to fuse these multi-channel EEG datasets by averaging, or implementing, another mathematical tool [2]. The independent control of multiple EEG channels cannot be handled when a non-specific alteration in the brain state occurs. However, it indicates significant correspondence with the mu rhythm frequency range (alpha range) [3]. As the scope of this device is only to trigger a switch, multiple EEG channels are not mandatory. The device can use only one channel to facilitate simplicity and cost effectiveness. If our model is used with single-channel EEG readings, we can get results, but the results will be less accurate than those received from a multi-channel system. Although the data processes are optimized, the results will still vary from person to person. The accuracy is in the range of 80% (±18%), even for relatively simpler tasks, indicating that the BCI technology is still not adequately ready for high-risk scenarios [4].

2.2 Review of Related Works

Persons who are paralyzed or have other severe movement disorders often need alternative means for communicating with and controlling their environments. Thus, BCI will play a crucial role. Many researchers discuss BCI and its applications. Among them is a study [3] where several persons learned to use two channels of bipolar EEG activity to control the 2D movement of a cursor on a computer screen. Although this independent control of two separate EEG channels cannot be attributed to a non-specific change in brain activity and appeared to be specific to the mu rhythm frequency range, with further development, multi-channel EEG-based communication may prove to be of significant value to those with severe motor disabilities. Meanwhile, another paper [5] describes real-time detection of brain events using EEG. Rather than working with off-line analogue recordings and datasets, it shows that it is possible to detect and classify evoked responses with surprising reliability. This paper has extensively deduced the solution of taking asynchronous, real-time readings. There is a brief overview of the research into BCI composing an offline study of the effect of MI on EEG and an online study that uses pattern classifiers incorporating parameter uncertainty and temporal information to discriminate between different cognitive tasks in real time [6].

Moreover, several papers [7,8,9] propose a BCI system that can bypass the normal motor output through the spine by using bioelectrical signals recorded on the intact scalp during purely mental activity, and this BCI model can categorize between brain waves for every voluntary action performed. This information is utilized for performing various functions such as moving robotic hands, controlling any prosthetic body part or, in this case, raising an alarm by triggering a functional device with the voluntary change of state of mind. The functioning of the brain is still a mystery, and reactions vary from person to person, even if slightly. However, this switch design tries to overcome that by mainly focusing on the MI of the person and hence on the change of frequencies in the brain signals. Every discovery in this field of technology will lead to a better and easier world for human beings. It needs mentioning that the main problem faced is that brain signals have extremely low amplitudes and therefore need a lot of amplifying before they can be used to perform any kind of task.

Borisoff et al. [10] propose an advanced concept of the real-life application of BCI to assist persons with motor disabilities. By comparing, recording and analyzing brain waves, BCI can help patients perform minor tasks using only decision making, which will be very helpful to persons with physical disabilities so that they can communicate with their nurses or family members. Padfield et al. [11] review the state-of-the-art signal processing technique for MI EEG-based BCIs. It mainly focuses on the prevalent challenges of implementation and commercialization of EEG-based BCIs. The BCI described in one paper [12] is an unsupervised algorithm to enhance P300-evoked potentials by the estimation of spatial filters. Then, the raw EEG signals are projected into the earlier estimated signal subspace. Neuper et al., in research work [13], performed an EEG-based feedback training program where patients were supervised from a distance using the telemonitoring system. Significant learning progress was found during the test with around 70% accuracy. An et al. [14] concentrate solely on the extraction and classification of EEG datasets based on MI tasks. Moreover, they applied deep learning algorithms. The aim of the research work by Heraz and Frasson [15] is to predict the three major dimensions of a person's emotions from brainwaves. The research investigates how machine learning techniques can be applied to determine the mental state of the test subject (pleasure/arousal/dominance). Standard classification techniques have been used to assess reliability. There are three popular signal processing techniques: (a) empirical mode decomposition, (b) discrete wavelet transform and (c) wavelet packet decomposition. Kevric and Subasi [16] focus on decomposition of EEG signals in the BCI system for a classification task, whereas Arvaneh et al. [1] have worked on improving BCI performance by removing irrelevant or noisy channels.

2.2.1 Different Brain Waves in the Brain

Numerous types of brain waves are emitted from different parts of the brain under multiple situations and scenarios. Brain waves of varying frequency ranges are emitted respective to the state of the brain. However, the primary problem with brain waves is that they exhibit very low amplitude. The external non-invasive EEG device that is used takes readings from the scalp of the subject, which results in lower effective amplitudes of the brainwaves. Also, the device often receives interferance from ambient noise, which increases complications while processing the data. Brain waves require a lot of signal processing to filter out the noise and artifacts. Moreover, if a multi-channel system is used, signal selection must be done to select the salient features from the extracted datasets. Then, we can identify what kind of waves are commanding the output signals and perform the corresponding action.

The different frequency bands of brain signals [15] are as follows:

- Delta waves: These waves are generated in the deepest meditation or dreamless sleep. They are associated with the process of healing. This is why sleep is so important in the process of healing.
- Theta waves: These waves are associated with learning, memory and, most importantly, intuition and knowledge that is beyond normal conscious awareness.
- Alpha waves: These waves are associated with flowing thoughts and some meditative states. When alpha waves are emitted, the brain is at rest. Alpha waves reflect overall calmness and relaxation.
- Beta waves: When attention is directed toward a cognitive task, beta waves are emitted. Beta waves are also emitted when we are alert, attentive, solving problems, making a decision or under stress or anxiety.
- Gamma waves: These waves can be used in invasive EEG devices, but they do not reach the scalp of the brain, so they cannot be implemented in non-invasive EEG devices. (Our switch model is based on a non-invasive EEG system.)

2.2.2 Electroencephalogram (EEG)

EEG devices are used to read brain wave activities using electrodes that are placed on the scalp of the subject. Brainwaves reaching the scalp are low in amplitude; hence, the signals are processed and amplified before they can be used for any computation. EEG signals have high time resolution; the device is more portable and relatively cost effective when compared to other technologies such as functional magnetic resonance imaging and magnetoencephalography (MEG) technologies [17,18]; hence, EEG is primarily used for BCIs. However, as EEG devices are portable, the signals are more vulnerable to external noise and artifacts [17,19]. Furthermore, the EEG signals can vary based on the body posture and emotional state of the subject [16].

2.2.2.1 Categories of EEG-Based BCIs

EEG-based BCIs are primarily categorized into two types: spontaneous cases and evoked cases [16]. In BCIs that use spontaneous control, the subject needs to voluntarily think of a specific action. The subject needs no external stimulation. This BCI will respond to the cerebral actions of the candidate. For example, if the person thinks of moving his or her arm, the BCI will respond to this mental activity and perform a task that has been pre-programmed [16,17]. In evoked BCI systems, the candidate, when subjected to an external stimulation, involuntarily triggers parts of the brain, and this change in the brain state is picked up by the BCI, which in turn performs a preprogrammed task [20].

2.2.2.2 Mechanism of the EEG

The EEG works with the help of a minimum of three electrodes. These three electrodes consist of one active electrode, a ground electrode and a reference electrode. The voltage difference between the active and the reference electrode gives us the data to plot a graph that helps us visualize the waves. The ground electrode is placed somewhere on the scalp, where it will not pick up ECG signals or any other electric signals from the body because that would cause interference in the brain waves, resulting in inconsistent results. The ground electrode completes the circuit so that the readings can be taken. The reference electrode is placed on the head at a position where the electrical signals are either too high or almost null. The active electrode is placed over the desired part of the head. The placement of the active electrode depends on the part of the brain from where we want to read the brain signals. There could be more than one active electrode in an EEG set, but all of them should be placed by the 10–20 rules.

2.2.2.3 Non-Invasive EEG Sensor Setup

Figure 2.1 shows the 10–20 rule scheme for placing electrodes for non-invasive EEG sensor setup. The 10–20 EEG electrode placement protocol has been standardized by the International Federation of Clinical Neurophysiology. The protocol uses the Nasion, Preauricular and Inion as reference points to divide the head into specific proportions, making sure all parts of the brain are covered. The diagram shows the electrodes are placed

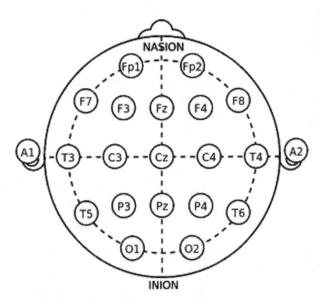

FIGURE 2.1
International 10–20 scheme of placing electrodes.

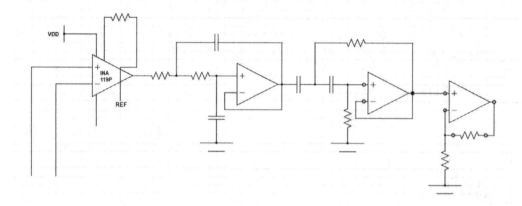

FIGURE 2.2
Circuit diagram of the EEG device.

all over the head with a distance of 10% or 20%. All the alphabets stand for a part of the brain (F for frontal lobe, T for temporal lobe, etc.). The numbers are the positions of the electrodes. Even numbers are used on the right side, and odd numbers are used on the left side.

2.2.2.4 Circuit Components of a Non-Invasive EEG Transducer

Figure 2.2 shows the circuit diagram of the EEG device. EEG-based BCI requires three steps of processing before we get raw data from the electrode readings:

a. Preprocessing
   In this step, we use an implementation amplifier to take the voltage difference between the active and the reference electrodes, and then we gain some amplitude in the output analogue signals. However, we cannot amplify the wave too much because of the noise that is still there.

b. Filtering
   The output from the instrumentation amplifier is passed through a low filter and then a high filter to exclude the ambient noise and the unwanted signals.

c. Amplification

Finally, the output wave is amplified to get an optimum amplitude that is usable for further operations.

2.3 Proposed Work: EEG-Based Simple Spontaneous Asynchronous BCI Switch (SSABS)

This proposed model demonstrates an asynchronous and spontaneous approach toward the BCI technology. The spontaneous property of this SSABS helps to assign the patient with maximum control of the device. This spontaneous characteristic of the device is implemented using MI, which is used to activate the alarm based on the voluntary EEG change of a target candidate. The alarm is also activated involuntarily when the candidate is in a state of panic. This is achieved by making the switch asynchronous and computing the data in real time. The asynchronous approach ensures that the data received from the EEG device is fed into the processor continuously and independently of the candidate's current brain state. Regardless of whether the person is in an active state (trying to trigger the switch) or in an idle state (doing some other activities), the readings from the EEG device are taken, constantly computed and checked for the given parameters. Solving cognitive tasks based on MI has a significant change in the brain signal readings of the candidate. When the mind is relaxed, it primarily emits alpha waves, and when the person tries to solve a problem, is alert or is in a state of panic, he or she emits beta waves. When the candidate voluntarily changes his or her state of mind from relaxation to a state of concentration (or trying to move, triggering the motor cortex), then he or she can use this change of brain state to activate the alarm. These features facilitate the versatile use and implementation of this model design.

2.3.1 Logical Components of the EEG Device Used in the BCI Switch

The BCI in this proposed model design is an EEG-based system, which means all the data collection from the candidate is conducted through EEG transducers. The logical components of the EEG are the basic compositional construct of our BCI design. Hence, it is essential to realize the basic components of the EEG system used and its ranges and parameters. These components have their respective significance and functions in the design model and collaborate to provide a useful output signal.

The following are the logical components of the EEG-based BCI switch.

- a. Brain: The brain signals are read with the help of electrodes placed over the head of the candidate. For example, if the candidate focuses on moving forward, then beta waves generate from the motor cortex of the brain. An electrode placed over the motor cortex reads the rise in beta waves. If the person is panicking, he or she experiences a sudden increase in beta three waves, which the electrode will detect.
- b. Instrumental amplifier: The switch takes two low-level signal inputs from the reference. The active electrode then amplifies the signal difference between the two inputs.
- c. Filters: The filter takes out the unwanted frequency that is not useful for further processing.
- d. Amplifier: The instrumentation amplifier cannot amplify the signal to the required amplitude because it still has unwanted frequencies. Hence, after filtering out the unwanted frequencies, a final amplification is required.

Figure 2.3 shows the components of the EEG-based BCI switch that have compositional significance. The active electrodes (one or multiple) and the reference electrode are connected to the instrumentation amplifier. The instrumentation amplifier amplifies the signal difference between the two (one active and the reference) electrodes and also rejects background noise to further pass the brain signals through the filter. A high pass filter and a low pass filter omit the unrequired range of frequencies. They pass on the usable part of the main brain signal. Alpha waves refer to the frequency range of 8–12 Hz,

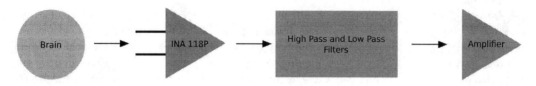

FIGURE 2.3
Components of the applied EEG device.

and beta waves correspond to brainwaves in the frequency range of 21–40 Hz, as expressed in Heraz and Frasson [15]. If we set the appropriate range of frequency, for example, a lower value of 8 Hz and upper value of 40 Hz, we will get a range where only alpha and beta waves can be received. Hence, we notice the change and dominance in the types of waves in this specific frequency band that will give us valuable information on setting parameters for the device. Then, these signals are amplified further with the help of an operational amplifier, and the gain is set to the required value with the help of resistors. The output signals are the raw datasets, which are then processed and computed to get an end result.

2.3.2 Workflow of the EEG-Based Simple Spontaneous Asynchronous BCI Switch (SSABS)

This section mainly focuses on explaining the detailed workflow of the SSABS in real-ife situations. In this theoretical implementation, we assume that the patient can perform any MI-based cognitive task. The patient must also be able to concentrate on the task. Figure 2.4 focuses on the detailed workflow of the EEG-based BCI switch in reference to its implementation in a real-life scenario.

First, a multi-channel or a single-channeled non-invasive EEG device is set up for the patient. If a multi-channel system is used, the 10–20 rule must be followed to place all the electrodes on the scalp of the candidate. However, for a single-channel system, the active electrode is placed over the motor cortex of the brain. This ensures that the data received from the low amplitude brain signal are of maximum clarity. The reference electrode is connected to the earlobe, where the brain wave activities are the lowest. The last electrode, the ground electrode, is placed behind the head. This completes the circuit that will henceforth provide us with the potential difference between the active and the reference electrodes as the output. The output signals are passed through the EEG device, and we get a slightly amplified dataset of our wanted frequency range (8–40 Hz consisting of only alpha and beta waves as explained in Heraz & Frasson [15]) along with reduced ambient noise. Using these raw datasets, we extract the features that are fundamental for our device to function. In the process, we also reduce the remaining unwanted noise and artifacts in our required frequency band that tend to reduce the resolution of the signals. This is referred to as feature extraction, and it is achieved by using signal processing techniques, namely time domain techniques, frequency domain techniques, time-frequency domain techniques or the CSP (common spatial space) technique. One of the most commonly used frequency domain feature extraction techniques is FFT (fast Fourier transform). FFT is used to obtain the power spectrum of the signals. From the power spectrum, we can easily determine the magnitude of the wave of the target frequency.

If a multi-channel EEG system is used, then the computational cost and complexity increases because then we will have datasets from multiple channels that need to be

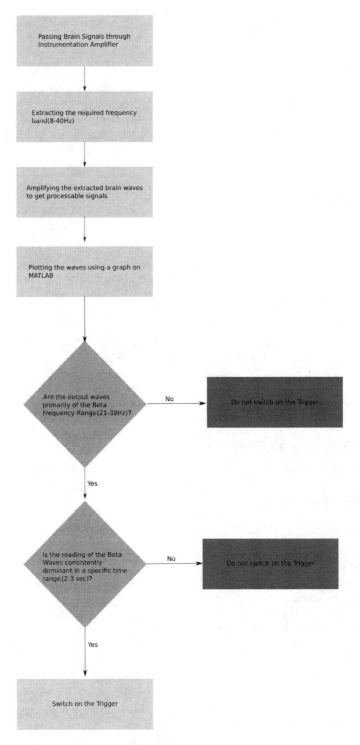

FIGURE 2.4
Workflow of the EEG-based BCI switch.

computed. Some of the channels might fail to read any useful data, and some channels may pick up high amounts of ambient noise. Then, we must employ a selection of valuable datasets to receive the final datasets that will not contain any unrequired or redundant data signals. Further, the average of the remaining datasets is taken to attain mean value readings. Further parameterized checking is done on these readings. Once the feature extraction and feature selection (only in multi-channel systems) is done, the data that are attained are amplified one last time, if required. The output waves are continuously checked in real time for any anomaly. Whenever a certain anomaly matches the parameters of the alarm system, the switch is triggered, and the alarm activates. In this case, we check for the anomalous spiking of beta waves. If beta waves are significantly commanding the output data signals, then the alarm is set off, indicating that the patient is calling for the attendant. However, if the subject exhibits a spike in beta three waves, then the alarm is activated, notifying the end user (nurse, attendant) that the patient is in a state of panic.

In a normal state of mind, the patient will show an approximate mu rhythm frequency band. The change between beta and mu rhythm frequency is the fundamental element of the theory on which this device is modeled. We also check that the beta waves are dominant in the output wave for more than a specific time range (i.e., 2–3 s). As per experimental data discussed in An et al. [14], classifiers for MI data tend to show inaccuracy after 2 s of mental activity. This happens because the MI tasks require relatively high concentration levels, and most of the candidates seem to lose concentration after 2 s. Hence, we have set a time segment of 2–3 s. This checking is mandatory because the brain signal activities are often very inconsistent when not used voluntarily. If the beta wave is not checked for dominant consistency in a specific time range, then the alarm could go off, raising a false alarm.

2.3.3 Salient Features of the Proposed Model

In this section, the most important aspects of our model are explained in order to exhibit the simplicity and versatility that this model design claims. The following are the salient features of the proposed model:

Spontaneous BCI System using MI (Motor Imagery)

The design of this switch is mainly based on the spontaneous system of BCI. The most widely acclaimed spontaneous BCI is MI BCI. BCI systems based on MI constantly examine the sensorimotor rhythms (SMR). These SMRs are oscillatory events that originate from the sections of the brain primarily associated with voluntary motion [11]. Implementation of MI in this BCI system will assure us that a large number of patients can perform this mental activity and use the device without any extensive training. According to past experiments, the motor cortex of the brain reacts to the change in alpha and beta waves in the EEG signals when an MI task is being executed by the candidate [18,22].

The activity in a specific frequency band of the EEG increases or decreases according to the mental activity of the patient. The increase in such activities is called event-related synchronization (ERS), and the decrease in activities in a previously more active frequency range is called event-related desynchronization (ERD). These ERDs and ERSs are triggered by MI activities and stimulation of senses [21,23,24]. Consider a patient with speech impairment or any kind of disability that makes it difficult to communicate with his or her attendant in conventional ways. If the patient uses this EEG-based BCI alarm system, he or she can voluntarily call the nurse or attendant by merely using his or her imagination (thinking of moving an arm or legs are the commonly used actions). This acts as the ERS in this case. The alarm rings for some specific but programmable time duration to notify the concerned person.

This BCI is an asynchronous system. By asynchronous, we mean that the readings will be taken throughout a course of time independent of the candidate's brain state. There will be an idle state where the patient is not trying to trigger the switch and another active state where the patient will voluntarily or involuntarily trigger the switch [10]. Active states can be realized during the moments when the subject is trying to perform any specific task that activates the motor cortex (imagining moving his or her left hand, etc.) or if the subject goes into a state of panic.

If commanding brain waves change from alpha waves to beta waves (the frequency of the brain waves change from a range of 8–12 Hz to a range of 12–38 Hz), then it is registered as an ERS and a trigger to raise the alarm. While in the idle state, the candidate may be doing other daily activities that fall under the scope of the patient's physical condition.

Asynchronous BCI System

Special Functionality to Identify a Panic State

This model will also be able to identify if the patient is having some kind of panic attack or facing any kind of high-energy emotions. Experimental research showed that when a subject is under panic or stress mainly beta waves are emitted from the brain [25]. Beta waves represent the frequency range of 13–40 Hz. However, when a subject panics, beta three waves (30–40 Hz), which are a sub-classification of beta waves, are emitted. After extraction of valuable segments from the raw data, we process it and classify the different frequency ranges of the beta waves. According to the final output datasets, we identify which frequency range of beta waves is dominating. We implement this information on the switch to indicate a panic attack or need of assistance and trigger the alarm, respectively.

2.4 Conclusion

The EEG-based BCI system functioning with the help of MI is complicated in theory and might have inconsistencies when tested on subjects. However, if implemented with adequate intelligibility and simplicity on a dynamic design model, it can do wonders. This chapter focuses on EEG-based SSABS, which works on MI using moto Rory excitations. This device has multiple components and subprocesses that hold significance. The design can be dynamic by making slight modifications to the parameters and device composition (i.e., changing some values in the filters or implementing a classifier on the end results can give varied but desired results, relevant to the specific modification). This technology has an infinite scope of implementation. More theoretical and experimental work will help to advance this technology so that it can have a significant effect on the lives of persons with disabilities.

The ultimate goal of this chapter is to implement a technological advancement that will revolutionize devices and instruments related to medical science. However, the technology can also be used in other fields such as gaming and entertainment to find innovative ways to give users a complex and higher-level experience. Once this BCI switch is implemented abundantly, medical monitoring of patients will change forever. Patients with physical disabilities, speech disabilities or partial paralysis will be able to communicate with their attendants without using conventional communication. The power of the human brain is unprecedented, and it is safe to say that we have not yet scratched the surface of its original potential. As of now, we are only able to identify changes in brain activities for voluntary and some involuntary actions performed by the subject. We might utilize this information for performing various functions such as moving robotic hands, controlling prosthetic body parts or raising alarms by triggering a functional device with the voluntary change of state of mind of the patient. The functioning of the brain is still a mystery, and reactions vary from person to person. However, this switch design tries to overcome that by focusing on the MI of the person and hence on the change of frequencies in the brain signals.

A more complex and updated version of this device would be a switch that implements a classifier to the end result datasets of the current model. In this way, we could classify the output waves into different categories. With enough experimental data and analysis it would be possible to discriminate between different MI (classes) (e.g., left hand, right hand, feet and tongue movement). So, if the different MI classes have some programmatic correspondence with different output tasks, then the scope of implementation of this switch increases drastically.

Every discovery in this field of technology will lead to a better and easier quality of life. However, future research must be conducted on how to get more accurate brain signal readings and how to reduce inconsistencies in readings of different test subjects. It is one of the most crucial aspects of improving this technology.

References

1. Arvaneh M. et al. (2011), Optimizing the channel selection and classification accuracy in EEG-based BCI. IEEE Transactions on Biomedical Engineering, vol. 58.6, pp. 1865–1873.
2. Padfield N., Zabalza J., Zhao H., Masero V., Ren J. (2019), EEG-based brain-computer interfaces using motor-imagery: Techniques and challenges. Sensors, vol. 19, no. 6, p. 1423.
3. Wolpaw J. R., McFarland D. J. (1994), Multichannel EEG-based brain-computer communication. Electroencephalography and Clinical Neurophysiology, vol. 90, no. 6, pp. 444–449.
4. Kim M., Kim M. K., Hwang M., Kim H. Y., Cho J., Kim S. P. (2019), Online home appliance control using EEG-based brain–computer interfaces. Electronics, vol. 8, no. 10, p. 1101.
5. Vidal J. J. (1977), Real-time detection of brain events in EEG. Proceedings of the IEEE, vol. 65, no. 5, pp. 633–641.
6. Penny W. D., Roberts S. J., Curran E. A., Stokes M. J. (2000), EEG-based communication: A pattern recognition approach. IEEE Transactions on Rehabilitation Engineering, vol. 8, no. 2, pp. 214–215.
7. Pfurtscheller G., Flotzinger D., Kalcher J. (1993), Brain-computer interface—A new communication device for handicapped persons. Journal of Microcomputer Applications, vol. 16, no. 3, pp. 293–299.

8. Wolpaw J. R., Birbaumer N., McFarland D. J., Pfurtscheller G., Vaughan T. M. (2002), Brain–computer interfaces for communication and control. Clinical Neurophysiology, vol. 113, no. 6, pp. 767–791.

9. Nicolas-Alonso L. F., Gomez-Gil J. (2012), Brain-computer interfaces, a review. Sensors, vol. 12, no. 2, pp. 1211–1279.

10. Borisoff J. F., et al. (2004), Brain-computer interface design for asynchronous control applications: Improvements to the LF-ASD asynchronous brain switch. IEEE Transactions on Biomedical Engineering, vol. 51.6, pp. 985–992.

11. Padfield N., et al. (2019), EEG-based brain-computer interfaces using motor-imagery: Techniques and challenges. Sensors, vol. 19.6, p. 1423.

12. Rivet B., et al. (2009), xDAWN algorithm to enhance evoked potentials: Application to the brain-computer interface. IEEE Transactions on Biomedical Engineering, vol. 56.8, pp. 2035–2043.

13. Neuper C., et al. (2003), Clinical application of an EEG-based brain-computer interface: A case study in a patient with severe motor impairment. Clinical Neurophysiology, vol. 114.3, pp. 399–409.

14. An, X., Kuang, D., Guo, X., Zhao, Y., He, L. (2014), A deep learning method for classification of EEG data based on motor imagery. In Proceedings of the International Conference on Intelligent Computing, Taiyuan, China, pp. 3–6.

15. Heraz A., Frasson C. (2007), Predicting the three major dimensions of the learner's emotions from brainwaves. International Journal of Computer Science, vol. 2, no. 3, pp. 187–193.

16. Kevric J., Subasi A. (2017), Comparison of signal decomposition methods in classification of EEG signals for motor-imagery BCI system. Biomedical Signal Processing and Control, vol. 31, pp. 398–406.

17. Van Steen M., Kristo G. (2015), Contribution to Roadmap. Accessed: 28 January 2019. Available: https://pdfs.semanticscholar.org/5cb4/11de3db4941d5c7ecfc19de8af9243fb63d5.pdf

18. Nicolas-Alonso L. F., Gomez-Gil J. (2012), Brain computer interfaces, a review. Sensors, vol. 12, no. 2, pp. 1211–1279.

19. Stawicki P., Gembler F., Rezeika A., Volosyak I. (2017), A novel hybrid mental spelling application based on eye tracking and SSVEP-based BCI. Brain Sciences, vol. 7, no. 4, 35.

20. Suefusa K., Tanaka T. (2017), A comparison study of visually stimulated brain–computer and eye-tracking interfaces. Journal of Neural Engineering, vol. 14, no. 3, p. 036009.

21. Pfurtscheller G., Aranibar A. (1977), Event-related cortical desynchronization detected by power measurements of scalp EEG. Electroencephalography and Clinical Neurophysiology, vol. 1977, no. 42, pp. 817–826.

22. Graimann B., Allison B., Pfurtscheller G. (2009), Brain-computer interfaces: A gentle introduction. In Brain-Computer Interfaces; Springer: Berlin, Germany, pp. 1–27.

23. Pfurtscheller G., Brunner C., Schlogl A. da Silva, F. H. L. (2006), Mu rhythm (de)synchronization and EEG single-trial classification of different motor imagery tasks. NeuroImage, vol. 31, pp. 153–159.

24. Leuthardt E. C., et al. (2004), A brain-computer interface using electrocorticographic signals in humans. Journal of Neural Engineering, vol. 1.2, p. 63.

25. Shanmuganathan D., Rajkumar R., Ganapathy V. (2018), Detection of panic and recovery from a panic using brain computer interface. International Journal of Pure and Applied Mathematics, vol. 119, no. 16, pp. 3655–3662.

3

Preprocessing and Discrimination of Cytopathological Images

Shouvik Chakraborty[1], Kalyani Mali[1], Sankhadeep Chatterjee[2] and Soumya Sen[3]

[1]*Department of Computer Science and Engineering, University of Kalyani, Kalyani, Nadia, West Bengal, India*
[2]*Department of Computer Science & Engineering, University of Engineering & Management, Kolkata, India*
[3]*A.K. Choudhury School of Information Technology, University of Calcutta, Kolkata, West Bengal, India*

CONTENTS

3.1 Introduction: Importance of Preprocessing

Microscopic images have some inherent characteristics that make preprocessing unavoidable. Raw data that comes from using a microscope may not be useful for different image processing algorithms. Preprocessing plays an important role in correct interpretation of the image data and mines useful information [1] using different computer algorithms [2,3]. Sometimes microscopic images contain very low spatial resolution that may affect the ultimate result of the analysis, which in turn causes the improper treatment of the patient. Microscopic images are also susceptible to noise. There are two basic types of noise that may contaminate the image. These are biological noise and noise due to the acquisition process [4–10]. Biological noises are inherent, and removal of this kind of noise is not easy. The image acquiring process may introduce some noise for several reasons. For example, improper illumination may cause poor image acquisition. The position of the different lenses in the microscope is one of the significant reasons for proper acquisition of the microscopic image. Preprocessing is essential for these images because huge deviation from the actual result is possible in the presence of noise [11]. Cellular images often contain different artifacts. These artifacts may cause improper segmentation and cause difficulty in detecting and analyzing the target regions or cells. Moreover, they may affect the cell counting algorithms. Automated systems may detect these artifacts incorrectly and increase the cell count, which in turn results in poor conclusions.

A clinical application needs accurate and precise results for correct diagnosis. Analysis of the microscopic images with the help automated systems may be difficult in the presence of different small objects and artifacts because it can influence the feature localization and detection methods [9,12]. Tracing the contours of different cells is an essential process that may not be efficient in the case of noisy images [8,13]. Some applications have been developed for microscopic and other biomedical images using semi-automated systems [14–17]. There are several issues related to this kind of application, too. Poor resolution, unclear edges and improper illumination make the process difficult. Moreover, semi-automatic processes may involve humans at certain stages that may introduce some inaccuracies. So, there are several causes behind the incorrect diagnosis of a microscopic image by automated systems, such as improper lights and illumination, shadows on some objects, noise, poor focus of the lens of the microscope, overlapped artifacts and objects, and similar nature of the background and the foreground. To overcome these issues, mathematical tools are needed to transform the image into the proper format so that it can be efficiently processed by computer-guided systems. Here, preprocessing methods come into the picture that can help to accurately analyze the microscopic image.

3.2 Enhancement

The main focus of the image enhancement methods is to process an image so that the image becomes more suitable for further processing. In general, different applications demand different types of images that better suit that application [18–23]. Enhance algorithms are designed and developed for a particular application. Enhancement methods try to enhance the image in different ways. For example, contrast of the image can be modified, and edges can be sharpened. One of the major challenges of the enhancement process is to find out appropriate criteria (mainly quantitative) for a specific modality and application. To overcome this issue, many applications use interactive methods for selecting appropriate criteria [9]. In general, image enhancement techniques can be categorized into two classes as follows:

  i. Spatial domain methods
 ii. Frequency domain methods

In the case of spatial domain analysis, the enhancement techniques work directly on pixels. One major advantage of this method is that it is very simple to study and requires less time, which helps to process data faster. These methods also support real-time applications [24] that require high processing speed. Sometimes these methods are not applicable due to poor robustness [25,26]. Spatial domain operations can be broadly classified into two general classes:

a. Point-based operations

b. Filter (spatial)-based operations

In frequency domain-based analysis, the original image is transformed into the frequency domain, and the image enhancement algorithms work on the coefficients of the transformed domain. There are several methods that can be applied to the actual image to transform it into the frequency domain. For example, Fourier transform [27], discrete wavelet transform [28], discrete cosine transform [29] and Haar wavelet transform [2,30–32]. The image is enhanced by modifying the coefficients obtained from the frequency domain data. Frequency domain operations provide the advantage of using some special features of frequency domain data. Moreover, having fewer computational steps makes it easier to use in several areas of application. Sometimes, frequency domain-based methods are unable to enhance all segments of the image, which makes it difficult to apply in some areas [26]. Frequency domain-based operations can be classified into three general classes, as given below [26].

a. Smoothing-based operations

b. Sharpening-based operations

c. Noise reduction-based operations

The image enhancement job can be thought of as a process to generate good quality images from poor quality images to make them suitable for a particular application. Image enhancement methods produce a better image in terms of both visual and quantitative measures, which helps in automated image analysis methods [4,9]. Analyzing both foreground and background information can reveal hidden information from the image that visual inspection does not reveal. In the case of color images, it is very difficult to extract objects from the foreground that have a similar color as the background and have poor edges. Microscopic image enhancement can be performed in two ways, as mentioned below.

a. Optical enhancement

b. Electronic/digital enhancement

Optical enhancements are a very useful and less costly method for enhancing a microscopic image. Some of them are darkfield, Rheinberg and polarized illumination. Some advanced microscopes use additional methods to get better enhancement. These methods may be less costly but very effective. Some of them are phase contrast, DIC and fluorescence [33,34].

Enhancement methods based on illumination are very effective and simple to implement. Sometimes a substantial amount of improvement can be achieved by simply changing the method of throwing light on the object under a test. That is the reason the methods are not expensive to implement. Some of the optical enhancement methods are discussed below.

- Darkfield illumination: In this method, a disk-shaped non-transparent object is inserted in the way of light so that the incident light looks like a hollow cone. This method can be used to explore very fine details. The disk-shaped object is generally termed a "stop" and can be a very simple thing such as a coin or paper.

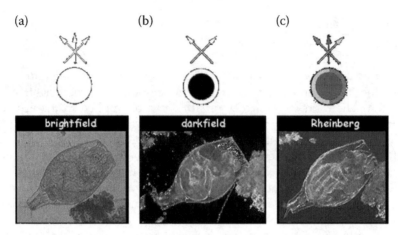

FIGURE 3.1
Example of (a) brightfield, (b) darkfield and (c) Rheinberg illumination.

- Rheinberg illumination: This method is similar to darkfield illumination but differs in the type of material used to design the stop. In this case, the stop is prepared with colored, translucent matter. The outer cone can be colored with different colors that will refract the light accordingly. The material used for the stop can be easily obtained from any stationary store. It can produce some interesting effects, as shown in Figure 3.1 [33,34].

- Polarized illumination: In this case, a pair of lenses is used to enhance the image. One lens is called a "polarizer", and the other one is an "analyzer". Various colors can be observed by rotating the polarizer. It is used in some of the sophisticated microscopes.

- Phase contrast: Contrast is one of the most important parameters for getting good quality microscopic images. In general, different parts of the cells absorb different amounts of light. This property is exploited to get better images. In the phase contrast method, a ring-like coating is used in one of the lenses. It is known as "phase annulus". A hollow cone-shaped light beam is used to illuminate the object.

- DIC: DIC stands for differential interference contrast. It is used in various applications and known for its better performance. It uses the polarizer along with the analyzer, similar to the polarized illumination method. However, it uses extra prisms. DIC can provide a 3D-like experience. Some "pseudo-colors" can be added to further enhance the image.

After obtaining an image from the microscope, it is possible to enhance it further using digital/electronic methods. Digital systems can enhance images by amplifying low light signals. They can control the brightness and contrast appropriately. Moreover, sharpness and some other features of the image can also be improved. Figure 3.2 [33,34] shows the effect of digital/electronic enhancement methods.

With the technological advancement, different techniques have been developed to cancel distortion and noise effects in the microscope. They can help to improve the quality of the image by eliminating various factors such as blur and noise, etc. [35–38]. Sometimes, three-dimensional information of an image can be used to perform better analysis and reconstruction. There are various sophisticated methods available that

(a)

(b)

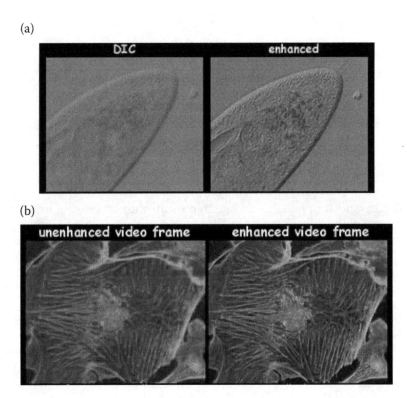

FIGURE 3.2
Effect of digital/electronic enhancement (a) DIC image (left) and increased apparent resolution (right) of a paramecium (b) fluorescence image (left) and enhanced image (right) of bovine pulmonary artery endothelial cell.

can be employed to achieve good results (some of them are of high mathematical complexity). Cellular biology has high dependency on these enhancement methods.

3.3 Different Categories of Cytopathological Images

Cytopathology deals with the cellular-level analysis of different diseases. It is considered to be a branch of pathology that is used to retrieve information from the cellular structure of living organisms. Cytopathology is generally concerned with the analysis of free cells or some parts of the tissue. It generally does not deal with the whole tissue. Cytopathology is not an old discipline [39] and has various scopes of research and improvement. Various disease analysis methods have high dependency on the cytopathological investigation. Different diseases, including cancer and various stages of infection, can be diagnosed efficiently and accurately, which gives insight about the status of the patient.

Cytopathological analysis is sometimes termed the "smear test" because the diagnosis process involves collection of a sample from different body parts, and then the sample is spread over a slide [19,35,36]. Sometimes it incorporates staining for microscopic analysis. In some cases, the sample can be prepared in other ways also.

To capture the digital image from the slides, the method that is used extensively is known as "whole slide imaging". It is basically a computer-aided method that is used to

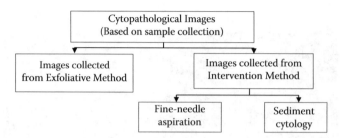

FIGURE 3.3
Classification of cytopathological images based on sample collection techniques.

digitize the glass slides to make them suitable for further processing using a computer [40,41]. This method is robust, is efficient and improves continuously. With the advancement of technology, whole slide imaging can be used to construct a three-dimensional view of the cells being tested [42–44]. Whole slide imaging technology is very powerful and helpful in finding a rough conclusion about the sample being tested. Moreover, it can handle a large number of slides simultaneously, which makes it more useful [42,45–48].

The field of cytopathology is experiencing a rapid growth, which makes it an affordable and reliable diagnostic tool. Manual investigation incorporates a huge amount of time and manpower, which can be reduced with the application of intelligent computer-aided and metaheuristic tools [38,49–54]. There are several methods of acquiring cytopathological images that can be further processed by a digital computer. On the basis of a sample collection strategy, cytopathological can be broadly classified into two classes, given in Figure 3.3.

With the exfoliative method, samples are collected from the peeled-off skin (without any external stimulus) or by mechanical means. In general, some specially designed spatula or brushes are used to collect the samples. Microscopic images are collected from these samples.

In the second method, the subject's body is intervened to collect the samples. The fine-needle aspiration-based method uses a needle to collect the image. A sediment cytology sample is collected from the external specimen that has been collected for some kind of test (e.g., biopsy). Microscopic images of the above-mentioned type are shown in Figure 3.4 [33,55–57].

Depending on the source (i.e., the part of the living organism from which the sample is being collected), the cytopathological images can be classified into various categories (e.g., gynecology-related samples, thyroid, eye, liver, breast, different glands). It is possible to analyze almost every cell collected from different parts of the body. Figure 3.5 [58–67] provides an illustration of different cytopathological image samples collected from different parts of the body.

3.4 Choice of Preprocessing Methods for Different Images

Image preprocessing is one of the essential tasks that should be performed for better analysis of cytopathological images. Preprocessing methods enhance an

(a)   (b)

(c)   (d)

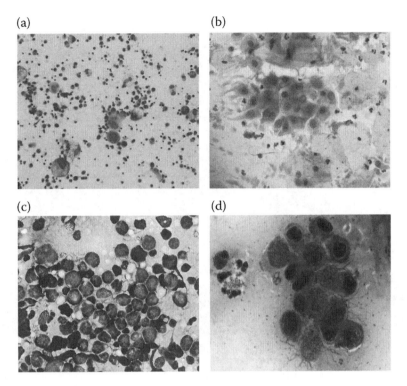

FIGURE 3.4
Different types of cytopathological images (based on sample collection strategy). (a) Spontaneous exfoliation, (b) mechanical exfoliation (pap smear), (c) intervention method – fine needle aspiration, (d) sediment cytology.

image by removing noise and unwanted artifacts to generate results with good accuracy and precision. Moreover, some special feature can be enhanced by some of the preprocessing methods that can be useful in further processing [19,35,36]. There are several methods available that can be applied to enhance a particular image. The choice of appropriate image preprocessing method is quite challenging because every image may not suffer from the same type of problem. Therefore, careful selection of preprocessing method is necessary; otherwise, poor result will be obtained [38,51,52]. Sometimes two or more methods are combined to get better results. In this section, some of the standard image preprocessing techniques will be discussed, along with their applicability and appropriateness for different types of cytopathological images.

Digital image preprocessing methods can be classified on several bases. For example, pixel-based transformations can be classified into four categories as follows:

   i. Operations on brightness
   ii. Geometric transformations
   iii. Restoration using the global knowledge
   iv. Local neighborhood-based operations

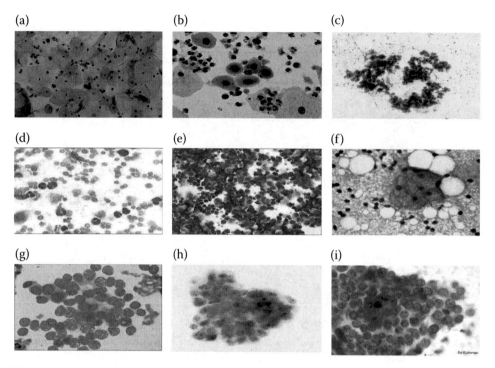

FIGURE 3.5
Different types of cytopathological images (based on the source section). (a) Cervical canal, (b) urinary tract, (c) breast (Syringoadenoma), (d) eye (conjunctival), (e) central nervous system (lymphoma), (f) salivary gland, (g) kidney (pyonephrosis), (h) respiratory (adenocarcinoma), (i) thyroid (papillary carcinoma).

In general, the overall image preprocessing methods can be outlined as follows:

- Enhancement
  a. Image filtering with spatial operations
     a. Point processing
        i. Contrast stretching
        ii. Image thresholding (global)
        iii. Histogram-based analysis
           - Equalization
           - Matching/specification
        iv. Log transform
        v. Power law transform
     b. Mask processing
        i. Smoothing (using filters, generally low pass)
        ii. Sharpening (using filters, generally high pass)
        iii. Median filters
        iv. Max filters

      v. Min Filters

      vi. Band/range filters

    c. Local thresholding

- Noise removal
- Detection and removal of skewness
- Size normalization
- Morphological operations
    a. Erosion
    b. Dilation
    c. Opening
    d. Closing
    e. Thinning and skeletonization
    f. Outlining

Before choosing any operation, we should gather some information. This information plays a vital role in determining the preprocessing method. Some of the key points are as follows:

- Knowledge about the nature/type of distortion.
- For microscopic cytopathological images, the nature of the microscope with the experimental conditions that have incorporated the noise (along with the type of noise).
- Segments or the objects for which we are interested should be taken care. It can significantly reduce the preprocessing workload. It is not always mandatory. Smart algorithms can acquire knowledge during the preprocessing phase.

Some standard methods for image preprocessing are as follows:

- Smoothing: The prime objective of the smoothing operation is to reduce noise. In general, a function is used to approximate the pixel value. In most of the cases, low pass filtering is used to smooth an image.
- Flat-field correction: Flat-field correction improves the quality of the image. This method is also known as background subtraction. Background with very low frequency is removed.
- Dilation: This is a morphological operation. The main target of this operation is to expand the connected regions. It investigates pixels around a window and makes a decision depending on a threshold.
- Erosion: This is a morphological operation that reduces the connected regions of an image. It investigates pixels around a window and makes a decision depending on a threshold.
- Opening: This is a morphological operation that is a combination of erosion and dilation. Dilation is performed, followed by erosion.
- Closing: This is a morphological operation that is a combination of erosion and dilation. Erosion is performed, followed by dilation.

- Basic filters: Some filters are used to improve the image quality. For example, max, min, median and mean. A window is considered for finding the new value.

Table 3.1 gives a brief overview about the types of cytopathological images on which the preprocessing methods can be applied.

Figure 3.6 [68–71] demonstrates some effects of the preprocessing techniques on different cytopathological images. Here, the impact of preprocessing can be visually justified. It can also be verified by using some quantitative methods.

TABLE 3.1

Choice of Preprocessing Methods Based on the Types of Cytopathological Images

Type of Issues Related to Cytopathological Images	Choice of Preprocessing Method
Noisy image	Smoothing, basic filters
Containing artifacts	Flat-field correction
Distortion due to the optical path of the microscope	Flat-field correction
Boundary discontinuity	Dilation
Intrusions	Dilation
Joined objects	Erosion
Extrusions	Erosion
Small foreground objects	Opening
Small holes in foreground	Closing

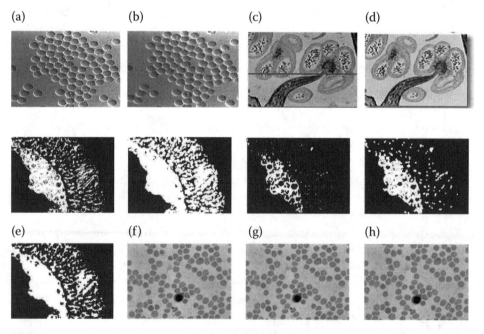

FIGURE 3.6

Effect of different preprocessing methods. (a) Original image of red blood cells; (b) smoothed image of red blood cells obtained from (a); (c) raw image; (d) flat-field corrected image obtained from (c); (e) original microscopic image of a cell; (f) after applying dilation on (e); (g) after applying erosion on (e); (h) after applying opening on (e); (i) after applying closing on (e); (j) original sample of blood smear; (k) after applying max filter on (j) and (l) after applying mean filter on (j).

3.5 Conclusion, Future Directions and Remarks

This chapter has discussed different methods of preprocessing that can be applied to cytopathological images. A brief overview has been presented on different types of cytopathological images. In general, preprocessing and enhancement are crucial steps that are essential for almost every application that deals with cytopathological images. As discussed earlier, the choice of the preprocessing method is highly subjective in nature and depends on the quality of image and the causes and sources for degradation. Careful acquisition of digital image reduces the distortion, which in turn reduces the overhead of preprocessing. Now, automated decision making about the selection of preprocessing method can be investigated further. The complexity of the preprocessing phase should be optimized so that the performance of various automated computer-aided real-time diagnostic systems improves. Enhancement procedures should not change the actual image for the sake of the processing simplicity; otherwise, the result of the diagnosis can be faulty, which can produce some drastic results in some situations because these systems are directly associated with the public health. Although lots of method have been developed to preprocess digital images, all methods are not foolproof, and they may not all perform well on different types of images. Moreover, due to some properties of cytopathological images (e.g., overlapping regions, presence of artifacts), it is always challenging to prepare them for further analysis. Research can be carried out in this domain to make the automated systems more powerful.

References

1. S. Datta *et al.* (2017), "Optimal Usage of Pessimistic Association Rules in Cost Effective Decision Making," in *2017 4th International Conference on Opto-Electronics and Applied Optics (Optronix)*, pp. 1–5.
2. K. Mali, S. Chakraborty, and M. Roy (2015), "A Study on Statistical Analysis and Security Evaluation Parameters in Image Encryption," *IJSRD-Int. J. Sci. Res. Dev.*, vol. 3, pp. 2321–2613.
3. K. Mali, S. Chakraborty, A. Seal, and M. Roy (2015), "An Efficient Image Cryptographic Algorithm based on Frequency Domain using Haar Wavelet Transform," *Int. J. Secur. Appl.*, vol. 9, no. 12, pp. 279–288.
4. S. Chakraborty, M. Roy, and S. Hore (2016), "A Study on Different Edge Detection Techniques in Digital Image Processing," in *Advances in Multimedia and Interactive Technologies - Feature Detectors and Motion Detection in Video Processing*, pp. 100–122. Available at: https://www.sciencegate.app/app/document/download/10.4018/978-1-5225-1025-3.ch005
5. S. Chakraborty *et al.* (2017), "Modified Cuckoo Search Algorithm in Microscopic Image Segmentation of Hippocampus," *Microsc. Res. Tech.*, vol. 80, no. 10, pp. 1051–1072.
6. S. Chakraborty, S. Chatterjee, A. S. Ashour, K. Mali, and N. Dey (2017), "Intelligent Computing in Medical Imaging: A Study," in *Advancements in Applied Metaheuristic Computing*, N. Dey, Ed. IGI Global, pp. 143–163.
7. S. Chakraborty, A. Raman, S. Sen, K. Mali, S. Chatterjee, and H. Hachimi (2019), "Contrast Optimization using Elitist Metaheuristic Optimization and Gradient Approximation for Biomedical Image Enhancement," in *2019 Amity International Conference on Artificial Intelligence (AICAI)*, pp. 712–717.

8. S. Hore *et al.* (2015), "Finding Contours of Hippocampus Brain Cell Using Microscopic Image Analysis," *J. Adv. Microsc. Res.*, vol. 10, no. 2, pp. 93–103.

9. S. Hore *et al.* (2016), "An integrated interactive technique for image segmentation using stack based seeded region growing and thresholding," *Int. J. Electr. Comput. Eng.*, vol. 6, no. 6.

10. S. Hore, S. Chatterjee, S. Chakraborty, and R. Kumar Shaw (2016), "Analysis of Different Feature Description Algorithm in object Recognition," in *Feature Detectors and Motion Detection in Video Processing*, IGI Global, pp. 66–99.

11. S. Chakraborty, A. Seal, M. Roy, and K. Mali (2016), "A Novel Lossless Image Encryption Method using DNA Substitution and Chaotic Logistic Map," *Int. J. Secur. Appl.*, vol. 10, no. 2, pp. 205–216.

12. S. Chakraborty *et al.* (2017), "An Integrated Method for Automated Biomedical Image Segmentation," in *2017 4th International Conference on Opto-Electronics and Applied Optics (Optronix)*, pp. 1–5.

13. S. Chakraborty *et al.* (2017), "Modified Cuckoo Search Algorithm in Microscopic Image Segmentation of Hippocampus," *Microsc. Res. Tech.*, vol. 80, no. 10, pp. 1–22.

14. H. E. Melton, S. M. Collins, and D. J. Skorton (1983), "Automatic Real-Time Endocardial Edge Detection in Two-Dimensional Echocardiography," *Ultrason. Imaging*, vol. 5, no. 4, pp. 300–307.

15. P. R. Detmer, G. Bashein, and R. W. Martin (1990), "Matched Filter Identification of Left-Ventricular Endocardial Borders in Transesophageal Echocardiograms," *IEEE Trans. Med. Imaging*, vol. 9, no. 4, pp. 396–404.

16. S. Hore *et al.* (2016), "An Integrated Interactive Technique for Image Segmentation using Stack based Seeded Region Growing and Thresholding," *Int. J. Electr. Comput. Eng.*, vol. 6, no. 6, pp. 2773–2780.

17. S. K. Setarehdan and J. J. Soraghan (1998), "Cardiac Left Ventricular Volume Changes Assessment by Long Axis Echocardiographical Image Processing," *IEE Proc. - Vision, Image, Signal Process.*, vol. 145, no. 3, p. 203.

18. S. Chakraborty and K. Mali (2018), "Application of Multiobjective Optimization Techniques in Biomedical Image Segmentation—A Study," in *Multi-Objective Optimization*, Singapore: Springer Singapore, pp. 181–194.

19. S. Chakraborty *et al.* (2018), "Dermatological Dffect of UV Rays Owing to Ozone Layer Depletion," in *2017 4th International Conference on Opto-Electronics and Applied Optics, Optronix 2017*, 2018, vol. 2018–Janua.

20. S. Chakraborty *et al.* (2018), "Bag-of-features based Classification of Dermoscopic Images," in *2017 4th International Conference on Opto-Electronics and Applied Optics, Optronix 2017*, 2018 a, vol. 2018–Janua.

21. S. Chakraborty, S. Chatterjee, A. Chatterjee, K. Mali, S. Goswami, and S. Sen (2018), "Automated Breast Cancer Identification by analyzing Histology Slides using Metaheuristic Supported Supervised Classification coupled with Bag-of-Features," in *2018 Fourth International Conference on Research in Computational Intelligence and Communication Networks (ICRCICN)*, pp. 81–86.

22. S. Chakraborty *et al.*(2018), "Bio-medical Image Enhancement using Hybrid Metaheuristiccoupled Soft Computing Tools," in *2017 IEEE 8th Annual Ubiquitous Computing, Electronics and Mobile Communication Conference, UEMCON 2017*, 2018, vol. 2018–Janua.

23. S. Chakraborty, S. Chatterjee, N. Dey, A. S. Ashour, and F. Shi (2018), "Gradient Approximation in Retinal Blood Vessel Segmentation," in *2017 4th IEEE Uttar Pradesh Section International Conference on Electrical, Computer and Electronics, UPCON 2017*, 2018, vol. 2018–Janua.

24. D. Sarddar, S. Chakraborty, and M. Roy (2015), "An Efficient Approach to Calculate Dynamic Time Quantum in Round Robin Algorithm for Efficient Load Balancing," *Int. J. Comput. Appl.*, vol. 123, no. 14, pp. 48–52.

25. M. Roy, K. Mali, S. Chatterjee, S. Chakraborty, R. Debnath, and S. Sen (2019), "A Study on the Applications of the Biomedical Image Encryption Methods for Secured Computer Aided Diagnostics," in *2019 Amity International Conference on Artificial Intelligence (AICAI)*, pp. 881–886.

26. S. S. Bedi, R. Khandelwal, A. Professor, and I. Bareilly (2013), "Various Image Enhancement Techniques-A Critical Review," *Int. J. Adv. Res. Comput. Commun. Eng.*, vol. 2, no. 3.

27. R. N. Bracewell (1989), "The Fourier Transform," *Sci. Am.*, vol. 260, no. 6, pp. 86–89.

28. M. J. Shensa (1992), "The Discrete Wavelet Transform: Wedding the À Trous and Mallat Algorithms," *IEEE Trans. Signal Process.*, vol. 40, no. 10, pp. 2464–2482.

29. N. Ahmed, T. Natarajan, and K. R. Rao (1974), "Discrete Cosine Transform," *Comput. IEEE Trans.*, vol. C-23, no. 1, pp. 90–93.

30. A. Seal, S. Chakraborty, and K. Mali (2017), "A New and Resilient Image Encryption Technique Based on Pixel Manipulation, Value Transformation and Visual Transformation Utilizing Single–Level Haar Wavelet Transform," in Proceedings of the First In*ternational Conference on Intelligent Computing and Communication*, Springer, Singapore, pp. 603–611.

31. A. Seal, S. Chakraborty, and K. Mali (2017), *A New and Resilient Image Encryption Technique based on Pixel Manipulation, Value Transformation and Visual Transformation Utilizing Single–Level Haar Wavelet Transform*, vol. 458.

32. R. S. Stankovir and B. J. Falkowski (2003), "The Haar Wavelet Transform: Its Status and Achievements," *Comput. Electr. Eng.*, vol. 29, no. 1, pp. 25–44.

33. J. A. Sullivan (1994), "https://www.cellsalive.com/enhance0.htm.", 1994. [Online]. Available: https://www.cellsalive.com/enhance0.htm. [Accessed: 10-Mar-2018].

34. J. A. Sullivan (1994), "https://www.cellsalive.com/." 1994. [Online]. Available: https://www.cellsalive.com/. [Accessed: 10-Mar-2018].

35. M. Roy *et al.* (2017), "Biomedical Image Enhancement based on Modified Cuckoo Search and Morphology," in *2017 8th Annual Industrial Automation and Electromechanical Engineering Conference (IEMECON)*, pp. 230–235.

36. M. Roy *et al.* (2017), "Cellular Image Processing using Morphological Analysis," in *2017 IEEE 8th Annual Ubiquitous Computing, Electronics and Mobile Communication Conference (UEMCON)*, 2017, pp. 237–241.

37. S. Chakraborty *et al.* (2017), "Image Based Skin Disease Detection using Hybrid Neural Network Coupled Bag-Of-Features," in *2017 IEEE 8th Annual Ubiquitous Computing, Electronics and Mobile Communication Conference (UEMCON)*, pp. 242–246.

38. S. Chakraborty, S. Chatterjee, N. Dey, A. S. Ashour, and F. Shi (2017), "Gradient Approximation in Retinal Blood Vessel Segmentation," in *2017 4th IEEE Uttar Pradesh Section International Conference on Electrical, Computer and Electronics (UPCON)*, pp. 618–623.

39. Kirkpatrick et al. (1989), *The Cassell Concise English Dictionary*. London: Weidenfeld Nicolson. p. 324. ISBN 0-304-31806-X.

40. J. Gilbertson and Y. Yagi (2008), "Histology, Imaging and New Diagnostic Work-flows in Pathology," *Diagn. Pathol.*, vol. 3, no. Suppl. 1. DOI: 10.1186/1746-1596-3-S1-S14

41. L. Pantanowitz (2010), "Digital Images and the Future of Digital Pathology," *J. Pathol. Inform.*, vol. 1, no. 1, p. 15.

42. F. R. Dee, A. Donnelly, S. Radio, T. Leaven, M. S. Zaleski, and C. Kreiter (2007), "Utility of 2-D and 3-D Virtual Microscopy in Cervical Cytology Education and Testing," *Acta Cytol.*, vol. 51, no. 4, pp. 523–529.

43. T. Kalinski *et al.* (2008), "Virtual 3D Microscopy using Multiplane Whole Slide Images In Diagnostic Pathology," *Am. J. Clin. Pathol.*, vol. 130, no. 2, pp. 259–264.

44. "(3) What kind of noise(s) are mostly present in microscopic image?" (2018) [Online]. (www.researchgate.net/post/what_kind_of_noises_are_mostly_present_in_microscopic_image. [Accessed: 12-Jan-2018].

45. D. C. Wilbur *et al.* (2009)., "Whole-slide Imaging Digital Pathology as a Platform for Teleconsultation: A Pilot Study using Paired Subspecialist Correlations," *Arch. Pathol. Lab. Med.*, vol. 133, no. 12, pp. 1949–1953.

46. J. Stewart, K. Miyazaki, K. Bevans-Wilkins, C. Ye, D. F. I. Kurtycz, S. M. Selvaggi (2007) "Virtual Microscopy for Cytology Proficiency Testing: Are We There Yet?," *Cancer*, vol. 111, no. 4, pp. 203–209.

47. M. C. Montalto (2008), "Pathology RE-imagined: The History of Digital Radiology and the Future of Anatomic Pathology," in *Archives of Pathology and Laboratory Medicine*, vol. 132, no. 5, pp. 764–765.

48. M. G. Rojo, G. B. García, C. P. Mateos, J. G. García, and M. C. Vicente (2006), "Critical Comparison of 31 Commercially Available Digital Slide Systems in Pathology," *Int. J. Surg. Pathol.*, vol. 14, no. 4, pp. 285–305.

49. S. Chakraborty and S. Bhowmik (2013), "Job Shop Scheduling using Simulated Annealing," in *First International Conference on Computation and Communication Advancement*, vol. 1, no. 1, pp. 69–73.

50. S. Chakraborty, A. Seal, and M. Roy (2015), "An Elitist Model for Obtaining Alignment of Multiple Sequences using Genetic Algorithm," in *2nd National Conference NCETAS 2015*, vol. 4, no. 9, pp. 61–67.

51. S. Chakraborty and S. Bhowmik (2015), "Blending Roulette Wheel Selection with Simulated Annealing for Job Shop Scheduling Problem," in *Michael Faraday IET International Summit 2015*, p. 100 (7.)–100 (7.).

52. S. Chakraborty and S. Bhowmik (2015), "An Efficient Approach to Job Shop Scheduling Problem using Simulated Annealing," *Int. J. Hybrid Inf. Technol.*, vol. 8, no. 11, pp. 273–284.

53. S. Chakraborty, M. Roy, and S. Hore (2016), "A Study on Different Edge Detection Techniques in Digital Image Processing," in *Feature Detectors and Motion Detection in Video Processing*, IGI Global, pp. 100–122.

54. S. Chakraborty *et al.*(2017), "Detection of Skin Disease Using Metaheuristic Supported Artificial Neural Networks," in *2017 8th Annual Industrial Automation and Electromechanical Engineering Conference (IEMECON)*, pp. 224–229.

55. David (2011), "PAP Smear." 2011. [Online]. Available: https://icytology.wordpress.com/the-pap-smear/. [Accessed: 12-Mar-2018].

56. J. T. Johnson (2018), "Fine-Needle Aspiration in Non-Hodgkin Lymphoma: Evaluation of Cell Size by Cytomorphology and Flow Cytometry." 2018. [Online]. Available: https://www.medscape.com/content/2002/00/43/80/438092/438092_fig.html. [Accessed: 12-Mar-2018].

57. S. Gupta, P. Sodhani, and S. Jain (2012), "Cytomorphological Profile Of Neoplastic Effusions: An Audit of 10 years with Emphasis on Uncommonly Encountered Malignancies," *J. Cancer Res. Ther.*, vol. 8, no. 4, p. 602.

58. B. J. Scott, V. C. Douglas, T. Tihan, J. L. Rubenstein, and S. A. Josephson (2013), "A Systematic Approach to the Diagnosis of Suspected Central Nervous System lymphoma," *JAMA Neurol.*, vol. 70, no. 3. pp. 311–319.

59. D. Rosenthal and S. S. Raab (2018), "Urinary Tract Cytology." [Online]. Available: https://icytology.wordpress.com/urinary-tract-cytology/. [Accessed: 12-Mar-2018].

60. F. Pacini and L. J. De Groot (2018), "Thyroid Nodules." [Online]. Available: http://www.thyroidmanager.org/chapter/thyroid-nodules/#toc-fine-needle-aspiration-cytology. [Accessed: 12-Mar-2018].

61. Heather kridel and Ruth Marion (2015), "Conjunctival Cytology: A Great Diagnostic Tool You're Probably Not Using!". 2015.[Online]. Available: http://blog.vetbloom.com/ophthalmology/conjunctival-cytology-a-great-diagnostic-tool-youre-probably-not-using/. [Accessed: 12-Mar-2018].

62. iCytology (2018), "CASE No 07 – Syringoadenoma of the Breast." [Online]. Available: https://icytology.wordpress.com/interesting-cases-2/case-n7-syringoadenoma-of-the-breast/. [Accessed: 12-Mar-2018].

63. Leonardo Da Vinci (2014), "Anatomy and Cytology of Salivary Glands." 2014. [Online]. Available: https://www.eurocytology.eu/en/course/302. [Accessed: 12-Mar-2018].

64. Microveda (2018), "Gynecology. Cervical canal. Cytological examination." [Online]. Available: https://www.youtube.com/watch?v=iJAFU7o5hvY. [Accessed: 12-Mar-2018].

65. N. Mangal, V. Sharma, N. Verma, A. Agarwal, S. Sharma, and S. Aneja (2009), "Ultrasound Guided Fine Needle Aspiration Cytology in the Diagnosis of Retroperitoneal Masses: A Study of 85 Cases," *J. Cytol.*, vol. 26, no. 3, p. 97.

66. PathologyApps.com (2018), "Adenocarcinoma Respiratory Cytology." [Online]. Available: http://pathologyapps.com/histology.php?id=489&type=cytology. [Accessed: 12-Mar-2018].

67. Pinterest (2018), "UA Sediment 6 - Atypical Cells and Dry Cytology Method." [Online]. Available: https://www.youtube.com/watch?v=572I2SExaD8. [Accessed: 12-Mar-2018].

68. D. K. Das, M. Ghosh, M. Pal, A. K. Maiti, and C. Chakraborty (2013), "Machine learning approach for automated screening of malaria parasite using light microscopic images," *Micron*, vol. 45, pp. 97–106.

69. IDL Online Help (2005), "Smoothing an Image." 2005. [Online]. Available: http://northstar-www.dartmouth.edu/doc/idl/html_6.2/Smoothing_an_Image.html. [Accessed: 13-Mar-2018].

70. K. R. Spring, John C. Russ, Michael W. Davidson (2018), "Basic Concepts in Digital Image Processing." [Online]. Available: https://www.olympus-lifescience.com/es/microscope-resource/primer/digitalimaging/imageprocessingintro/. [Accessed: 13-Mar-2018].

71. S. E. Umbaugh (1990), *Digital Image Processing and Analysis: Human and Computer Vision Applications with CVIPtools*, 2nd Edition, Boca Raton, FL: CRC Press, Taylor & Francis Group, 2011, p. 956, ISBN: 9781439802052.

4

Applications of Internet of Vehicles in the Medical Sector

Kanishka Sharma[1] and Jayanta Mondal[2]

[1]*Department of CSE, Mody University of Science and Technology, Rajasthan, India*
[2]*School of Computer Engineering, KIIT University, Bhubaneswar, India*

CONTENTS

4.1 Introduction

With rapid development of the social economy and the process of industrialization, the number of vehicles in cities has increased dramatically. However, owing to limited capacity, roads have become saturated, which causes considerable issues, such as traffic congestion, traffic accidents, energy consumption and environmental pollution. In the process of exploring solutions to the above-mentioned issues, the automotive industry introduced new energy sources and energy-saving vehicles to achieve the coordinated development of the automotive industry and environment. In addition, it is essential to fully use the information and communication technologies for achieving the coordinated development of human, vehicle and environment, which can alleviate traffic congestion, enhance transportation efficiency and enhance existing road capacity. Internet of Vehicles (IoV) emerged and quickly became the research focus, both in academia and the industry. Owing to the various unsolved issues, it is significantly challenging to implement IoV. The predominant challenges are described as follows:

1. Inconsistent understanding of IoV hinders its development. Owing to different people understanding IoV differently, people from different fields propose different concepts and architectures of IoV. However, these differences lead to problems such as human and vehicle coordination problems, which impede the development of IoV.

2. Human, vehicle and environment should be integrated in IoV. With the development of the intelligent vehicle, vehicles have become an addition to human perception. Hence, human, vehicle and environment integration is the core of IoV service.

3. The coordination behaviors between human and vehicle are considerably complex. Because the vehicle movement behaviors exhibit an impact on human decisions, intelligent vehicles cannot simply act as humans and make decisions. Humans and vehicles should coordinate to make decisions. The IoV application scenarios are spatiotemporally diverse, which leads to the complexity of human and vehicle coordination.

To overcome the above challenges, this chapter first focuses on the concept and architecture of IoV. The definition of IoV is provided, and the architecture of IoV based on the functional requirements and expected goals is proposed. The architecture includes the environment sensing and control layer, the network access and transport layer, the coordinative computing control layer, and the application layer (Figure 4.1).

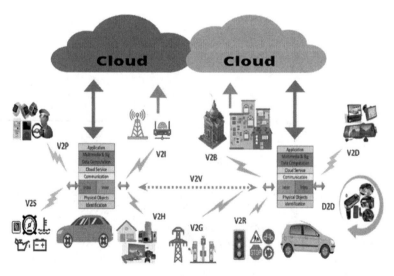

FIGURE 4.1
General architecture of IoV.

4.1.1 Concept and Definition of IoV

Our cities are becoming smarter and more connected, and this allows connected vehicles to slowly transform into one autonomous unit, but none of this will be possible without a new advanced network. A key element that is among the reasons for such rapid growth in the use of IoT devices is the IoV. It allows vehicles to exchange information efficiency and most importantly safety with others as well as with infrastructures using vehicular ad hoc networks (VANETs), which originated from MANET, or the mobile ad hoc network. IoV is the evolution of that conventional VANET, which refers to the network of different entities such as vehicles, pedestrians, roads, parking lots and city infrastructure, and it provides real-time communication among them.

The electronics used in this include infotainment systems, sensors, brakes and GPS. There is a clear requirement for better communication and interconnectivity between vehicles. Because they are turning into smart entities, cars are becoming an essential part of smart cities. IoV makes car sensor platforms take information from the environment, vehicles and drivers for safer navigation, traffic management and pollution control.

The IoV is a network of cars communicating with each other and with pedestrians' handheld devices, roadside units (RSUs) and public networks using vehicle-to-vehicle (V2V), vehicle-to-road (V2R), vehicle-to-human (V2H) and vehicle-to-sensor (V2S) interconnectivity. This creates a network with intelligent devices as participants. As an important branch of IoT in the transportation field, IoV covers a wide range of technologies and applications, including intelligent transportation, vehicular information service, modern information and communications technology, and automotive electronics. However, owing to different understandings of the connotation of IoV in various research fields, there is no uniform definition of IoV. Some related definitions are provided in the following. IoV is considered to be a vehicle-connected network that is based on a vehicular information system. Drivers are informed of the status of their vehicle based on outside information that is sensed by the vehicle's electronic equipment. The essence of the IoV study is based on the drivers. Intelligent transport systems (ITS) realizes the traffic guidance and control through the interconnection between vehicle and vehicle, as well as the vehicle and the environment, which essentially studies IoV from the perspective of traffic.

i. Alam et al. regard IoV as an ITS-integrated IoT from the perspective of smart transportation. In other words, IoV is the vehicular ITS that regards the driving vehicle as the information-sensing object. IoV integrates the intelligent sensors, wireless communication technology, distributed data, information processing and internet technology for achieving the efficient exchange of information and sharing involving human and vehicle, vehicle and vehicle, and vehicle and roadside infrastructure.

ii. From the viewpoint of network interconnection, Hartenstein et al. consider that IoV is a network that integrates the in-vehicle network, inter-vehicles network and vehicular mobile internet. They found that radio-frequency identification (RFID), GPS, mobile communication, wireless network and network service support technology in IoV are used for achieving the real-time wireless communication and information exchange system network between vehicle and road, vehicle and vehicle, vehicle and human, and vehicle and cities.

iii. Li presents the definition of IoV from the perspective of integration of on-board sensors and communication technology. They consider IoV to be vehicles carrying advanced sensors, controllers, actuators and other devices, which integrate modern communications and network technology for providing the vehicles' complex environmental sensing, intelligence decision making and control functions.

4.2 Architecture of IoV

Some researchers propose the architecture of IoV based on IoT, which is similar to IoV and consists of the sensing layer, network layer and application layer. However, IoV is not only a service network for vehicle-to-vehicle communication or vehicle terminals but also a complex system that has the feature of human-vehicle-environment with tightly coordinative interaction and highly dynamic evolution. In the process of cooperation among human, vehicle and environment, IoV is required for supporting pervasive computing, cognitive computing and new types of information computing, which requires novel requirements on architecture and support capacity for IoV. To attain this target, we integrate the architectures of IoV proposed by existing academia and industry, and we split it into four layers, that is, vehicle network environment sensing and control layer, network access and transport layer, coordinative computing control layer and application layer.

Vehicle Network Environment Sensing and Control Layer

Environment sensing is the acknowledgment basis for IoV services, such as services of autonomous vehicle, intelligent traffic and vehicle information. The control of the vehicle and the traffic environment object are the basis of IoV services implementation.

From the perspective of vehicles, they sense environment information around themselves via the autopilot system, traffic jam auxiliary structure and sensor system for achieving auxiliary driving. In terms of the environment, this layer studies the procedure to monitor and extract dynamic information of humans, vehicles and the environment through sensing technology. In addition, it focuses on the procedure to receive and execute coordinative control instructions and then feedback results to co-operative control. Finally, it implements the capabilities of swarm sensing in the swarm model and participates in sensing using the individual model.

Network Access and Transport Layer

The basic idea of this layer is realizing inter-connection and information exchange, which includes the access network, the transmission network and the control network. The access network provides real-time, three-dimensional and seamless heterogeneous network access for vehicles connecting to the network. The transmission and control network responds to dispatch access resources and balances information load. Then, it establishes a stable, quality-guaranteed information and communication transmission channel considering the network load conditions and access resource constraints.

Coordination Computing Control Layer

This layer provides IoV applications with the network-wide capability of coordinative computing and control for human-vehicle-environment, such as data processing, resource allocation and swarm intelligence computing. From the perspective of IoV objects' coordinative individual model, this layer should provide the capability of human-vehicle coordinative computing control to support the human-vehicle, which achieves coordination in the IoV environment. From the perspective of IoV objects' coordinative swarm model, this layer should provide the capabilities, including the multi-human and multi-vehicle coordinative computing control and service coordinative management, for the purpose of supporting the swarm intelligence computation and various services. In addition, in order to meet the coordinative control requirements, this layer should also provide the capability of communication coordinated management.

Application Layer

The application layer provides various types of services to achieve the requirements of human-vehicle-environment coordination services. In addition, the application layer should be in the open state and should share information to support the novel service and business operating model. Owing to the closed status quo of the application, the application layer can be classified into closed services and open services.

The closed services are closely related to the specific industry applications, such as the intelligent traffic command and control platform. The open services cover various existing open applications, such as real-time traffic service provided by various internet service providers. The application layer should also provide open service capabilities to third-party providers (Figure 4.2).

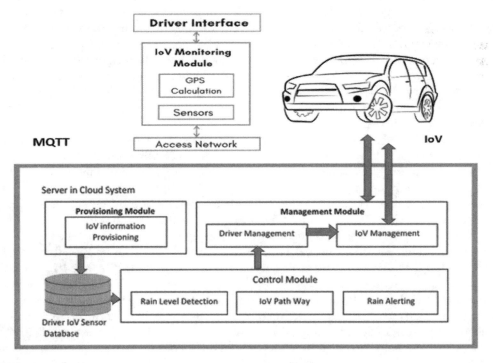

FIGURE 4.2
System architecture of IoV.

4.3 IoV Coordination Computing Based on Virtual Vehicle

At present, the studies of IoV predominantly focus on micro and macro levels. At the micro level, the research focuses on intelligent vehicle driving and driving safety, which makes micro decisions based on the surrounding environment. These decisions invariably do not consider the impact of traffic. Therefore, IoV cannot solve the existing traffic problems effectively. From the macro level, the research focuses on the procedure to guide and control vehicle flow, which makes macro decisions based on the global traffic information and the current traffic state. However, the macro decisions do not consider the personalized behavior of the vehicle. Moreover, the solution invariably indicates inefficiency when it is used for solving the traffic problem.

Moreover, because the traffic prediction in macro studies depends on historical traffic flow, the results are not accurate. To eliminate the gaps from the middle studies level in IoV, we propose configuring an image that has intelligence of vehicle and driver in cyberspace, named Virtual Vehicle (VV).

VVs can connect individual vehicle control and vehicle flow control from the middle level, and realize the human-vehicle-environment coordination control. A VV is the image in cyberspace (e.g., cloud) of the human and vehicle in physical space. This image includes the features and characteristics of human and vehicle. VV embodies the microcosmic behavioral features of driver and vehicle during the driving. Moreover, VVs can interact directly with each other in cyberspace by providing traffic services and sharing sensing data coordinately, which can solve the bottleneck of communication in physical space.

A VV combines the driverless vehicle and ITS from the middle perspective. Moreover, we provide the architecture of a VV. The architecture includes the perception and execution layer in physical space and the supporting environment and service layer in cyberspace. The perception and execution layer provides the ability to sense and control the human, vehicle and environment in physical space. The same driver in different vehicles can compose different VVs, and the same vehicle with different drivers can also compose different VVs. In this layer, the driver-vehicle cognition model should be constructed accordingly to sense the driving behavior and vehicle characteristics. This model not only senses the information of driver and vehicle, but also recognizes the differences of driving behavior with different drivers.

Based on the differences, we can obtain the personalized characteristics of a VV, which provides the personal services and the basis of swarm coordination control.

The supporting environment, which is provided by the coordination computing control layer, provides a platform for VV operation, and computing capability and public service capability. A VV interacts with the corresponding vehicle through the vehicle-to-network communication, which can realize the coordination between a VV and a driverless vehicle. Simultaneously, VVs can interact with each other in the environment. Therefore, they can realize the swarm intelligence computation. In addition, it is essential for the supporting environment to deal with the traffic data from ITS, the sensing and interaction data by itself, which are the basis for a VV in decision making and interaction. In the supporting environment, the computing capability and public service capability can be provided for a VV to compete its computing and service function. The service layer includes on-board services and traffic services. On-board services include vehicle navigation services, vehicle communication services and vehicle entertainment services, such as telematics. Traffic service predominantly provides traffic control and environment-coordinated control, and it pushes accurate traffic services to the traffic control department or other transportation service providers. For the on-board service, a VV can provide personalized services according to the results of global traffic information and interaction with others.

For traffic service, the traffic control department can obtain the vehicle's current state and predict future behavior accurately because of the existence of VVs in cyberspace, which can be used for realizing the message broadcast and traffic control.

4.4 Related Works

Recently, various attempts have been made to survey the existing body of work proposed to solve the security and privacy issues in the connected vehicles arena.

First, some of the surveys at this point are now considered outdated, such as work in the La and Cavalli surveys up to 2017. Second, some surveys are more threat-centric, meaning that the authors focused more on representing the severity of the threats than on countermeasures. However, some of them follow a solution-centric approach; that is, they focus more on describing the result spectrum than the threats posed. In this, we use a systematic literature review (SLR) in order to study the existing proposed work related to securing connected vehicles. We use SLR to present comprehensive and unbiased information regarding various state-of-the-art security problems, solutions and proposals in VANETs and IoVs. We briefly discuss these approaches as follows:

1. Engoulou et al. surveyed the security issues and the challenges in VANETs and also introduced a variety of architectures to address the security issues. The authors mostly focused on security problems and threats. The proposed work cited in the paper only goes to 2017, with no mention of an internet of connected vehicles.

2. Similarly, the recent work in Contreras et al. is related to IoV in that the authors discussed IoV protocols, architectures and standards, but did not look comprehensively at security. A more recent work in this direction that focused more on threats is Eiza and Ni, who focused on cybersecurity threats, such as malware, auto mobile app-related threats, and on-board diagnostic (OBD) vulnerabilities, and also discussed the countermeasures proposed for them.

3. The authors in Othmane et al., proposed categorization of security and privacy aspects for IoV. The authors named these aspects as information validity, device security, communication links' security, identity and liability, privacy and access control. The authors surveyed the projected schemes according to their taxonomy. The authors in Parkinson et al. also categorized and surveyed the work connected to securing vehicles from cyber threats.

4.5 Benefits of IoV

In this section, we examine benefits of IoV and compare IoV with traditional vehicular communication (i.e., VANETs). Secondly, we evaluate IoV in the context of two new paradigms in vehicular communication; namely, cyber-physical system (CPS) and named data networking (NDN).

4.5.1 IoV over VANETs

A comparative investigation between IoV and VANETs is carried out to highlight the fruitful impact of the realization of IoV on vehicular communication, its services and the business orientation of the communication.

The following observations can be made from this comparative assessment:

- The vehicular communications of IoV would be highly commercialized. This is due to the smart commercial and infotainment applications in addition to the smart safety, management and efficiency applications.
- The network architecture of IoV would integrate vehicular communication with other communication networks. This is due to the various network architectures.
- IoV provides a reliable internet service in vehicles. This is due to the inclusion of V2I communication.
- Most of the existing computing and communication devices would be compatible with vehicular networks of IoV.
- The processing and decision-making capability of vehicles, the size of vehicular networks and the volume of network data would enlarge drastically in IoV.

4.6 Challenges and Issues

The complete realization of IoV could bring fundamental changes to the driving experience by integrating smartness into ITS applications. Apart from driving, other related areas, including urban traffic management, automobile production, repair and vehicle insurance, road infrastructure construction and repair, logistics and transportation, would be positively transformed. The cost effectiveness in these areas would be a major contribution of IoV. This is due to the collaborative integration of available information for optimizing cost. However, comprehending IoV would be a significantly challenging task. Various issues of vehicular communications are yet to be resolved for the realization of IoV. Some of challenges are described below considering their specific issues.

4.6.1 Localization Accuracy

Accurate localization of vehicles is significantly challenging considering the accuracy requirement in vehicular communication environments. The accuracy requirement is higher than the accuracy provided by existing GPS-based localization. In fulfilling the accuracy requirements, the following three issues need to be addressed.

- GPS-based localization provides accuracy of 5 m whereas the accuracy requirement in vehicular communication environments is 50 cm.

- GPS-based localization does not take the speed of the objects into consideration whereas speed is one of the important constraints in vehicular communication environments.
- Deteriorated quality of GPS signal and unreachability of signal is a concern in dense urban environments.

4.6.2 Location Privacy

Due to the highly mobile ad-hoc network environment, vehicular communications are constructed by periodically beaconing information about the network. The periodically beaconing information comprises location, velocity, direction, acceleration and vehicle type. Revealing location information can be a huge privacy concern. The vehicles have to utilize location information for communication without revealing the information itself. Therefore, location privacy is a challenging task. Although some techniques, including pseudonyms, silent periods and mix zones, have been suggested to address the privacy concern, the concern remains unresolved. This is due to the list of issues below.

- Pseudonym switching is feasible in the case of higher vehicle density. The technique is easily detectable in environments with less vehicle density.
- A silent period is relevant for non-real-time ITS applications. The procedure is not suitable for real-time applications.
- A mix zone is suitable for multi-lane roads with larger zone areas. The method is not effective on one-way roads.

4.6.3 Location Verification

Location verification of neighboring vehicles is another challenging difficulty in vehicular communication. This is due to the lack of trusted authorities in vehicular communication. Some of the techniques recommended for location verification include directional antenna, beaconing-based belief and cooperative approach. The following issues need to be addressed in these techniques:

- The cost of infrastructure in the directional antenna approach and the limitations of range-based techniques in the vehicular environment
- The overload involved in the beaconing approach
- The untrustworthy neighbor in cooperative verification

4.6.4 Radio Propagation Model

Radio propagation in the vehicular traffic environment is significantly deteriorating due to the modern road infrastructure and speed of vehicles. The radio obstacles on and alongside roads can be categorized into moving and static radio obstacles. The moving obstacles on the road include trucks, buses and other larger size vehicles. The static obstacles include buildings alongside roads, flyovers, underpasses and tunnels. Currently, wireless propagation models of mobile communications are used in vehicular communications; thus, we do not consider the impact of the aforementioned obstacles on the radio propagation in vehicular environments. Moreover, the WAVE standard for vehicular communication uses 5.9 GHz frequency, which has smaller penetration capability compared to well-known WiFi and mobile radio signals. Therefore, the following concerns need to be addressed in order to develop accurate radio propagation models for vehicular communications.

- Incorporation of the impact of moving and static obstacles in radio propagation models for vehicular communications.
- Maintaining accurate line of sight in vehicular communications considering the lesser penetration capability of 5.9 GHz vehicular radiofrequency.

4.6.5 Operational Management

Due to the collaboration and coordination among different types of networks, the volume of data in IoV would drastically increase. Therefore, operational management in terms of security and credibility would be considerably challenging. Vehicles would be equipped with a number of sensors, different types of radio terminals and transponders.

As a result, a multi-attribute network would emerge where the traditional network operators such as Internet Service Provider, telecom operators, automobile companies and dealers would collaborate and work under a third-party virtual network operator. Apart from the equipment-associated complexities, computational complexities would also be challenging due to the contemplation of cloud computing in highly mobile network environments. Apart from the mentioned challenges and issues, there are other issues related to disruption reduction, opportunistic framework, geographic routing and media access control standard.

The issues have been explored up to some extent in the context of the existing vehicular communication architecture. Thorough investigations are needed in the context of the framework of IoV.

4.7 Applications

Apart from using it in the medical sector, IoV has diverse applications, which include the following:

i. Safe driving: This refers to cooperative collision avoidance systems that use sensors to detect imminent collision and provide warning to the driver. This application involves periodical status messages and emergency messages. An emergency message is generated by an emergency event such as a traffic jam, an accident or a bad road condition.

ii. Traffic control: IoV will bring about fundamental alterations to urban congestion management, transport and logistics, urban traffic and our collective lifestyle.

iii. Crash response: Connected cars can automatically send real-time data about a crash along with a vehicle's location to emergency teams. This can save lives by accelerating the emergency response.

iv. Convenience services: IoV provides the ability to remotely access a car. This enables services such as remote door lock and recovering a stolen vehicle or "find my vehicle", which is extremely helpful if you have left your vehicle in a huge parking lot. Furthermore, the technology is very helpful for transportation agencies as it improves real-time traffic, transit and parking data, making it easier to manage transportation systems for reduced traffic and congestion. IoV is also responsible for advanced infotainment systems. Connected cars can provide online, in-vehicle entertainment options that enable streaming music, media, navigation or any other information through the dashboard.

v. Infotainment: Connected cars can offer online, in-vehicle entertainment options that provide streaming music and information through the dashboard.

4.8 Conclusion

In IoV, human, machine, vehicle and environment can synchronize. Mobile internet, ITS, cloud computing, automotive electronics and intelligent geographic information systems are integrated via advanced information communication and processing technology. Hence, the industry chain of IoV includes the automotive manufacturing and services, communications and information services, financial and insurance services, and public service fields. This chapter reviews the development of IoV from the perspective of architecture and key technology and summarizes the existing challenges in IoV. Then, the chapter configures the VVs in IoV in cyberspace and provides the nature and architecture of a VV [1-8].

References

1. Lu, Z., Liu, W., Wang, Q. (2018), A privacy-preserving trust model based on blockchainfor VANETs. *IEEE Access*; vol. 6, pp. 45655– 45664.
2. Contreras, J., Zeadally, S., Guerrero- Ibanez, J. A. (2017). Internet of vehicles: architecture, protocols, and security. *IEEE Internet Things J*; vol. 5, pp. 3701– 3709.
3. Eiza, M. H., Ni, Q. (2017). Driving with sharks: rethinking connected vehicles with vehicle cybersecurity. *IEEE Vehic Tech Magazine*; vol. 12, pp. 45–51.
4. Sakiz, F., Sen, S. (2017). A survey of attacks and detection mechanisms on intelligent transportation systems: VANETs and IoV. *Ad Hoc Networks*; vol. 61, pp. 33–50.
5. Butt, T. A., Iqbal, R., Shah, S. C. (2018). Social Internet of vehicles: architecture and enabling. *Comp ElectrEng*; vol. 69, pp. 68–84.
6. Parkinson, S., Ward, P., Wilson, K. (2017). Cyber threats facing autonomous and connected vehicles: future challenges. *IEEE Trans Intel Transport Syst*; vol. 18, pp. 2898–2915.
7. Upadhyaya, A. N., Shah, J. S. (2018). Attacks on VANET security. *Int J Comp Eng Tech*; vol. 9, no. 1, pp. 8–19
8. De La Torre, P., Rad, G., Choo, K.-K. R. (2018). Driverless vehicle security: challenges and future research opportunities. *Future Generat Comp Syst*. Epub ahead of print 11 January 2018.

5

CNN-Based Melanoma Detection with Denoising Autoencoder for Content-Based Image Classification

Dhritiman Mukherjee

Amity School of Engineering and Technology,
Amity University Kolkata, Kolkata, India

CONTENTS

5.1 Introduction

Content-based image classification can be considered as one of the most efficient choices for computer-aided diagnosis in recent time for data-driven decision making in computer science. It assists patients and medical teams by choosing the correct line of treatment by analyzing correctly. Melanoma is one the biggest concerns in recent times because of its effects on society. One of the vital reasons of its negative effects is late or delayed detection. Melanoma is widespread in Europe, North America and Australia, and there has been a tremendous increase in cases of late detection of melanoma in the last 5 years.

On the other hand, far greater availability of smartphones featuring multi-core CPU and sensors have the potential to enable people to monitor their own health problems.

DOI: 10.1201/9780429318078-5

Our contribution is designing the cheap smartphone-based model that can allow users to test skin abnormalities in their home, office, etc., in real time. An ordinary smartphone is used for this project as a platform. Thus, the proposed system is easily accessible.

As will be further discussed, the proposed system is based on a client server architecture. There are basically two components. The first component is a smartphone platform where lightweight images are captured, and the second component is a background server where classification algorithms are running.

Detection of melanoma using smartphone-captured images is quite new. Most of the previous work focused on dermoscopic images on clinical equipment. These types of clinical treatment sometimes are costly or are inaccessible for people living in remote locations.

Content-based image retrieval (CBIR) uses feature extraction. Ultimately, feature extraction begins from an initial set of measured data and construct-derived values (features) that are intended to be insightful and non-redundant, enabling the subsequent learning and generalization steps, and in some cases, leading to better human interpretations.

Researchers place the feature extraction technique into two groups, namely hand-crafted descriptor extraction and automatically learned descriptors. Hand-crafted descriptors are extracted low-level characteristics of image data such as shape, color and texture. Examples are histograms of color, histograms of oriented gradients, scale-invariant feature transforms (SIFT) and speeded up robust features (SURF). A color histogram is a depiction of the color distribution in an image to describe color. Histogram-oriented gradients (HOG) is a feature descriptor used to define shapes.

The second group is automatically learned descriptors. A deep neural network algorithm was successfully introduced for identifying abnormalities in demographic images. Convolution neural network (CNN) used for the purposes of image classification can extract image neural code. These neural codes are the features that are used to describe the images. Feature extraction using CNN demonstrates state-of-the-art approaches on many datasets. The problem with this procedure, however, is that labeled data are needed to train the neural network. The labeling task can be costly and time consuming.

To solve this problem and to generate these neural codes, an unsupervised deep learning algorithm is used by means of a denoising autoencoder (DAE). A DAE is a feedforward neural network that learns to denoise images and interesting features on the images used to train it. An overall flow diagram is shown later in Figure 5.1.

5.2 Related Work

The DAE introduced by Pascal et al. [1] as an extension of the classic autoencoder is a recent addition to the image denoising literature. Pascal et al. [2] showed that the DAE can be stacked to form a deep neural network. In their proposed work, Jain et al. [3] showed image denoising using a convolution network.

Recently, mobile-enabled dermatoscopic devices have been developed such as DermLite and HandyScope; however, the cost of such devices is very high, and accessibility is limited.

Furthermore, trained personnel need to handle such devices.

There are a few works that utilize smartphone sensors and cameras. This type of application is very cheap and easily accessible. Many systems work on mobile platforms, and many transmit images to the local server. However, users need to consider the image noise or distortion effect while taking pictures or sending the image to the remote server. For example, a few studies [4,5] use a mobile dermatoscope to capture dermatoscopic images and send them to a server for assessment.

There are a few isolated works that use the mobile platform for computation, capturing images from the smartphone itself. In Ramlakhan and Shang [6], mobile platform-based detection is presented. They used a very basic thresholding method to detect a lesion. Doukas et al. [7] used a mobile platform and remote cloud for lesion detection and classification. Lesion detection and feature extraction were computed locally inside the mobile device, and the back-end server was used for image classification purposes. However, the authors concentrated on system integration.

Even though recent work [8,9] proposes a complete architecture, the effectiveness of this system running on limited resources is not presented. Recently, there are many DNN systems available for image analysis, such as lesion segmentation [10] and melanoma detection [7,11–16]. However, due to resource constraints such as memory and computation, they are highly challenging to use on a smartphone platform.

5.3 Methodology

Figure 5.1 shows the overall architecture. Initially images are captured on devices such as smartphones. Query images are then preprocessed because images may be subject to various types of distortion. Smartphone-captured images for melanoma skin cancer detection pose some challenges, but the problem is not well understood [1,2]. Preprocessed images are then supplied to the next level of the system to extract the region of interest (ROI) containing the lesion. Identification of ROI greatly reduces the size of the feature, with the least impact on precision.

The ROI contains images that are then fitted into CNN for feature extraction. The extracted feature is evaluated with classifiers, namely artificial neural network (ANN) to predict the image type.

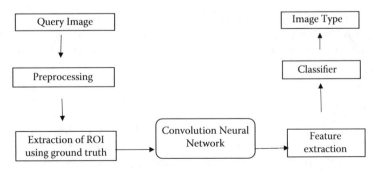

FIGURE 5.1
Architecture of proposed methodology.

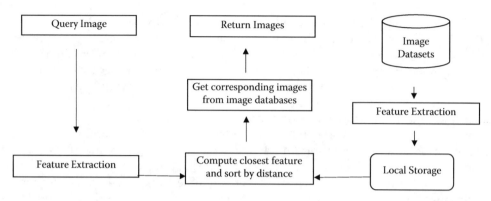

FIGURE 5.2
Architecture of the process of denoising.

5.4 Denoising Technique

Image noise is a very common problem that can affect industry and scientific applications. Images are corrupted with noise in every acquisition channel. The noise can be harmful to many image applications. Hence, it is very important to remove the noise from images. Denoising by the neural network is a current research area [3]. The overall architecture of the process of denoising is shown in Figure 5.2. Our project uses a CBIR system with a DAE, which can efficiently remove noise and find different images by querying images between image datasets. Images are indexed in CBIR by their visual content such as color, texture and shape. The key point about CBIR is feature extraction.

The dataset is taken from the ISIS (International Skin Image Collaboration) archive. It consists of 1,800 pictures of benign moles and 1,497 pictures of malignant classified moles. Initially, the program extracts features from each image of the dataset and saves them into the local database. Then, the program extracts features from the query images. Next, the program applies a Euclidian distance technique to sort by distance to compute the closest features. Corresponding images from the dataset have been sorted based on the closest features.

5.4.1 Autoencoders

Encoding and decoding can be carried out to minimize the error rate by means of a neural network known as an autoencoder [4]. An autoencoder takes an input $x \in [0, 1]^d$ and then maps it with an encoder to the hidden layer $y \in [0, 1]^{d'}$ through deterministic mapping, e.g.,

$$z = \sigma(Wx + b) \tag{5.1}$$

where σ can be considered to be non-linear such as the sigmoid. The latent representation y or code is then mapped back with a decoder into a reconstruction z of the same shape as x. This mapping happens through the similar transformation, e.g.,

$$x' = \sigma'(W'y + b') \tag{5.2}$$

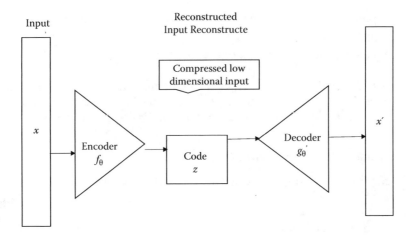

FIGURE 5.3
Autoencoder model architecture.

where the prime symbol in (5.2) does not indicate it is a matrix transpose. Accessing the given code y, x' can be seen as the prediction of x, W is a weight matrix and b is a bias vector. Weights and bias vectors are randomly assigned, and then modified and updated through back propagation with iteration during training. After the encoding stage, the decoder stage of the autoencoder maps z to reconstruct x' in the same shape as x.

The parameters of the model (W, W', b, b') are optimized to minimize the reconstruction error. Reconstruction error can be measured by means of squared error and cross-entropy. Reconstruction error is often referred to as a loss. The traditional square error $L(x, x') = ||x - x'||^2$ can be used if the input is a bit vector.

$$L(x, x') = ||x - x'||^2 = ||x - \sigma'(W'(\sigma(Wx + b)) + b')||^2 \qquad (5.3)$$

The basic architecture of the autoencoder is shown in Figure 5.3. The input layer is encoded with z in one hidden layer, and the output is reconstructed at the output layer. The encoder network mostly accomplishes the dimensional reduction. The decoder layer takes a compressed and low dimensional representation of the input, very similar to principal component analysis.

The model contains an encoder function f(.), parameterized by θ and a decoder function g(.) parameterized by θ. The low-dimensional code learned for input x in the bottleneck layer is z, and the reconstructed input is $x' = g_\theta'(f_\theta(x))$

5.4.2 Denoising Autoencoder (DAE)

The DAE is a stochastic version of the autoencoder. The DAE first tries to encode the input preserving the information about the input, and second it tries to undo the effect of a corruption process stochastically applied to the input of the autoencoder.

The steps of training DAE works are as follows:

- The data are partly distorted by applying noises or masking stochastically certain values of the input vector, $\tilde{x} \sim M_D(\tilde{x} | x)$. [$\tilde{x} = x + \text{noise}$].

- The corrupted input $\tilde{x}$ is then mapped to a hidden representation of the standard autoencoder with the same operation, $z = f_\theta(\tilde{x}) = \sigma(Wx + b)$.
- At the final step, the hidden representation model is reconstructed $x' = g_\theta'(h)$.

The model's parameters θ and θ' are trained to minimize the average reconstruction error over the training data, specifically minimizing the difference between z and the original uncorrupted input x. The basic architecture of the DAE is shown in Figures 5.3–5.6.

5.4.3 Convolutional Neural Network (CNN)

The name "convolutional neural network" indicates that the network is associated with a special type of mathematical operation called a convolution, a specialized kind of linear operation. The basic architecture of a CNN consists of an input layer, an output layer and several hidden layers. The hidden layers of CNN consist of a series of convolution layers. The activation function is commonly a RELU layer and is subsequently followed by additional convolutions such as pooling layers, fully connected layers and normalization layers, referred to as hidden layers because their

FIGURE 5.4
Original input image x; partially destroyed image $\tilde{x}$.

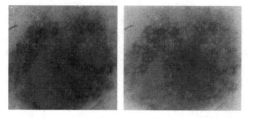

FIGURE 5.5
Denoising autoencoder with some inputs set to missing.

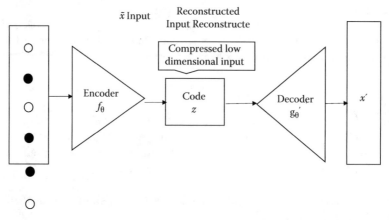

FIGURE 5.6
Denoising autoencoder model architecture.

inputs and outputs are masked by the activation function and final convolution. CNN is the most successful model for supervised classification and is state of the art in many benchmarks [5,6].

5.4.4 Convolution Autoencoder (CAE)

Convolution autoencoders (CAEs) are types of CNNs and are state-of-the-art tools for convolution filters. They are based on the traditional autoencoder architecture with encoder and decoder layers. Once the filters have learned to extract features, they can be applied to any input for feature extraction. This feature can be used further for input such as classification. Every convolution filter is wrapped with an activation function a to improve the generalization capability.

5.4.5 Encoder

A convolution among input volume $I = \{I_1, \ldots, I_D\}$ and set of n convolution filter $\{F_1^{(1)}, \ldots, F_n^{(1)}\}$ each of depth D produces a set of n activation maps, a volume of activation maps with depth n:

$$Z_m = O_m = a\left(I_* F_m^{(1)} + b_m^{(1)}\right) \quad m = 1, \ldots, m \quad (5.4)$$

where $b_m^{(1)}$ is bias for the mth feature map. Z_m is used for latent variable name in the autoencoder.

5.4.6 Decoder

The produced n feature is a map for the input of the decoder in order to reconstruct the input image I from the reduced representation. The reconstructed image $\tilde{I}$ is the convolutional output between feature Z_m and filter volume $F^{(2)}$.

$$\tilde{I} = a\left(Z_* F_m^{(2)} + b^{(2)}\right) \quad (5.5)$$

where $b^{(2)}$ is the bias per input channel.

5.5 Result and Discussion

This research work was carried out with the core i3 system with Samsung smartphones with 13 MP cameras. This system was the basic architecture of the client server. The client part was designed in Android and tested on Samsung smartphones with 13 MP cameras. The server part was designed in Python and tested on a core i3 system with 8 GB of RAM. Figure 5.7 shows the classification results of first few moles in the dataset.

The dataset is taken from the ISIC archive, which consists of 18,00 images of benign moles and 1,497 images of malignant classified moles. The dataset is pretty balanced, and it has two different classes of skin cancer (i.e., benign and malignant). In this chapter, 64 filters are used for the first two Conv2D layers.

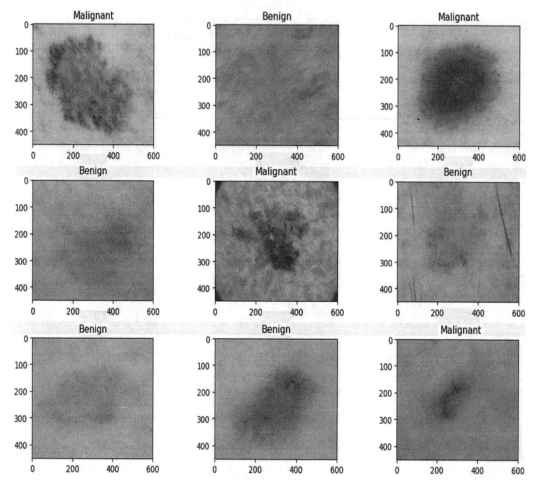

FIGURE 5.7
Displaying first nine images of moles and how they classified.

Kerastensor flow in the background was used to build the CNN model for this project to detect different classes of moles. The first important layer in CNN is the convolution (Conv2D) layer. The Conv2D layer can be imagined as a set of learnable filters.

The second layer of the CNN is the pooling (MaxPool2D) layer. This layer basically acts like a down sampling filter. These are used to reduce overfitting and computational cost. By combining the convolutional and pooling layers, CNN is able to learn global feature by collecting local features. RELU acts as a rectifier. The rectifier activation function is used in the network to add linearity to the network.

5.6 Conclusion

Melanoma is a big concern in recent times and is a significant health hazard that needs early detection to cure. This chapter discussed a system for malignancy detection by means of CNN and DAE. Images taken from the mobile device are used to diagnose cancer by means of an ISIC dataset. In our future work, we would focus on high-resolution images by implementing other denoising methods such as singular value decomposition median filter before image preprocessing.

References

1. P. Vincent, et al. (2008), "Extracting and composing robust features with denoising autoencoders." In *Proceedings of the25th international conference on machine learning.* ACM.
2. P. Vincent, et al. (2010), "Stacked denoising autoencoders: Learning useful representations in a deep network with a local denoising criterion." *Journal of Machine Learning Research* 11: 3371–3408.
3. K. Ramlakhan and Y. Shang, (2011), "A mobile automated skin lesion classification system," In *ICTAI '11: Proceedings of the 2011 IEEE 23rd International Conference on Tools with Artificial Intelligence,* November, pp. 138–141. https://doi.org/10.1109/ICTAI.2011.29
4. Y. Gu and J. Tang (2014), "A mobile system for skin cancer diagnosis and monitoring," *SPIE Mobile Multimedia/Image Processing, Security, and Applications.*
5. O. Abuzaghleh, M. Faezipour, and B. D. Barkana (2015), "Skincure: An innovative smart phone-based application to assist in melanoma early detection and prevention," *Signal & Image Processing : An International Journal (SIPIJ),* vol. 5, no. 6, pp. 1–15. DOI: 10.5121/sipij.2 014.5601
6. K. Ramlakhan and Y. Shang (2011), "A mobile automated skin lesion classification system,"In *IEEE 23rd International Conference on Tools with Artificial Intelligence, ICTAI 2011,* Boca Raton, FL, USA, November 7–9, DOI: 10.1109/ICTAI.2011.29.
7. C . Doukas, P. Stagkopoulos, C. T. Kiranoudis, and I. Maglogiannis (2012), "Automated skin lesion assessment using mobile technologies and cloud platforms," *Annual International Conference of the IEEE Engineering in Medicine and Biology Society,* 2012, pp. 2444–2447.
8. T.-T. Do, Y. Zhou, V. Pomponiu, N.-M. Cheung, and D. C. I. Koh (2017), "Method and device for analyzing an image," Patent US 20 170 231 550 A1, August 17, 2017.
9. T. T. K. Munia, M. N. Alam, J. Neubert, and R. Fazel-Rezai (2017), "Automatic diagnosis of melanoma using linear and nonlinear features from digital image." In *39th Annual International Conference of the IEEE Engineering in Medicine and Biology Society (EMBC).*
10. M. H. Jafari, E. Nasr-Esfahani, N. Karimi, S. M. R. Soroushmehr, S. Samavi, and K. Najarian (2017), "Extraction of skin lesions from nondermoscopic images for surgical excision of melanoma," *International Journal of Computer Assisted Radiology and Surgery,* vol. 12, no. 6, pp. 1021–1030.
11. V. Pomponiu, H. Nejati, and N.-M. Cheung (2016), "Deepmole: Deep neural networks for skin mole lesion classification." In *IEEE International Conference on Image Processing (ICIP).* IEEE.
12. A. Esteva, B. Kuprel, R. A. Novoa, J. Ko, S. M. Swetter, H. M. Blau, and S. Thrun (2017), "Dermatologist-level classification of skin cancer with deep neural networks," *Nature,* vol. 542, pp. 115–118.
13. A. Krizhevsky, I. Sutskever and G. Hinton (2017), "ImageNet classification with deep convolutional neural networks," *Communications of the ACM,* vol. 60, no. 6, pp. 84–90.

14. W. Wang, Y. Huang, Y. Wang, and L. Wang (2014) "Generalized autoencoder: A neural network framework for dimensionality reduction." In *Proceedings of the IEEE conference on computervision and pattern recognition workshops*, pp 490–497.

15. A. Krizhevsky (2009), "Learning multiple layers of features from tiny images." Master's thesis, Computer Science Department, University of Toronto.

16. Y. LeCun, L. Bottou, Y. Bengio, and P. Haffner (1998), "Gradient-based learning applied to document recognition," *Proceedings of the IEEE* 86(11), pp. 2278–2324.

6

IoT and Fog Computing-Based Smart Irrigation System for the Planting Medicinal Plants

Anil Saroliya[1], Ananya Singh[2] and Jayanta Mondal[3]

[1]*Department of Information Technology & Engineering, Amity University, Tashkent, Uzbekistan*
[2]*School of Engineering and Technology, Mody University of Science and Technology, Rajasthan, India*
[3]*School of Computer Engineering, KIIT University, Bhubaneswar, Odisha, India*

CONTENTS

6.1 Introduction

Medicinal plants can be used to develop herbal medicines and many other related products. These plants are helpful to develop and sustain cultural values worldwide. Some plants are a significant source of nutrition; therefore, they are recommended for remedial use to cure many diseases. Plants such as aloe vera, rosemary, basil and thyme have high medicinal value and require little water to grow. In this chapter, we discuss the smart irrigation system for medicinal plants that can be cultivated in desert areas. As water resources are scarce and continuously depleting, smart ways of irrigation are a necessity. This work defines how to handle irrigation intelligently with the use of Internet of Things (IoT) and fog computing. The main thing which has been targeted in this work is to save the farmers time, money and energy as well as to avoid problems such as

DOI: 10.1201/9780429318078-6

overflowing water and the need to keep vigilant. The work also aims to provide efficient irrigation techniques to medicinal plants that require less water to grow. Therefore, this project is highly beneficial for such medicinal plants. Not only the plants mentioned above but also plants having similar medicinal value can be irrigated with this project. It also aims to give solutions to real-world problems for water conservation by automatically supplying water to needed areas as required in plantation fields. This system could prove to be useful in large agricultural lands as well as small backyard gardens. The objective is to detect the soil moisture contents of the surroundings and, based on the sensing, provide water through the water pump module. The information regarding the soil moisture, surrounding temperature and humidity will be sent to the Thing Speak Server for graphical representation and also to the user's mobile phone using Blynk application.

Agriculture has been a crucial part of life since the prehistoric ages as it provides food and other raw materials. The growth of the agricultural sector of a country is very important for its economic advancement. The agricultural sector of India largely contributes to the GDP of the country [1]. Therefore, smart irrigation techniques are needed to improve the pre-existing agricultural growth rate for future development. It is unfortunate that most farmers in India still use traditional, manually operated techniques for irrigation. The traditional irrigation system includes overhead sprinklers, drip irrigators, etc., which require consistent vigilance and also waste water resources [2]. These frameworks also encourage flood-type irrigation, which typically wets the lower leaves and stems of the crops or plants. The whole soil surface is immersed and remains wet long after the water supply is finished [3]. Such a condition advances diseases in plants and leaves by organisms such as fungi [3]. These antiquated techniques can be replaced with new automated techniques. This chapter focuses primarily on reducing the manual labor involved in maintaining the fields and also on ways to reduce waste of water. The recent advances in soil water monitoring systems along with the growing universality of fog computing and wireless sensor networks make the commercial use of this system applicable for large-scale agriculture and small-scale farms as well [4]. The system is designed to irrigate or supply water at regular intervals of time depending on the soil moisture contents [2]. In this system, there are three major parts involved: the humidity sensing part, a control section and an output section. The humidity sensing is performed by the temperature and humidity sensor and also the soil moisture sensors. The control section is controlled by the NodeMCU platform, and the output is the data displayed on the Thing Speak Server in the form of a graph, on the user's mobile phone in the form of notifications and also on the serial monitor. Also based on the data comparisons water is regulated from the water tank to the plants through the water pump module. In this technique, the soil sensors are placed near the root portion of the plants; they sense the humidity in soil and send the data periodically to the Thing Speak Server for graphical representation. The NodeMCU module handles the sensor data and compares it with a predefined threshold value, which is programmed in the NodeMCU board through the Arduino IDE. The user also gets notifications on his or her smartphone through the Blynk application as to whether the motor regulating water to the plants is turned on or off. The user can also access the sensor data on his or her mobile phone provided the user stays connected to the internet.

There are several herbal and medicinal plants that require very little water for growth and can grow with the help of this irrigation method. These plants include aloe vera (*Aloe barbadensis miller*), rosemary (*Salvia rosmarinus*), thyme (*Thymus vulgaris*) and basil (*Ocimum basilicum*) [5]. Other such plants are turmeric, ginger, mint and cilantro [6].

There are a number of other plants that are helpful and can be irrigated through this process for their medicinal value.

Aloe vera is an evergreen perennial that originates from the Arabian Peninsula. It is most commonly found in tropical, semi-tropical and arid climates. The benefits of this plant are endless. It is beneficial for teeth and gums, constipation, diabetes-induced foot ulcers, antioxidant and possible antimicrobial properties, protection from UV irradiation and protection from skin damage after radiation therapy [5].

Rosemary is also a commonly found herb, and it is sometimes used to make a number of herbal teas. It is high in antioxidants and anti-inflammatory compounds, and it helps to lower blood sugar level, boost mood, improve memory, support brain health and protect vision and eye health [7].

Thyme is used to lower blood pressure, to stop coughing, to boost immunity, to get rid of pests, and to boost mood and mental health.

Basil also has many medicinal uses such as reducing oxidative stress, supporting liver health, fighting cancer, protecting against skin aging, reducing high blood sugar, reducing inflammation and swelling, and combating infection [6].

These are a few of the medicinal plants that can be irrigated using the proposed system. This system is therefore beneficial for both agricultural and medicinal purposes.

The proposed system irrigates fields based on soil dampness and humidity in the surroundings and at the same time updates the status of irrigation wirelessly through a server and an application. It will allow farmers to continuously monitor the moisture level in the cultivation area and control the water supply remotely through the internet, thus achieving an optimal level of irrigation using the IoT.

6.2 Literature Survey

Naturally, all flora require water to survive. However, some do not need consistent availability of water and are called drought tolerant plants. They are the category of plants that can survive months or even years of scarce rainfall due to their adaptation. With the variations in weather patterns and conditions across different terrestrial regions, plants in drought areas have adapted to survive the harsh climatic conditions of thriving without water [5].

With the advancement in technology, there is an evident need for more efficient and specialized systems for better irrigation. Various research has been carried out to make irrigation more efficient. The researchers have used different ideas and technologies depending upon the soil condition and water quantity to design models for effective and efficient irrigation. Similar systems have been proposed by researchers using the Arduino board as well [8]. The principle motive behind designing this system is to ease human labor. The initial stage involves placing the soil moisture sensors directly into the root zone near the plant. The soil sensors sense the water or moisture level of the ground, and if that sensed level is inadequate or below the predefined threshold value, then the control unit commands the water pump module to supply water to the required area. The specific purpose of the control unit platform is to control the various connections and interactions between the modules, enabling them to sense and smartly act on-site [9]. The basic components for this competent and automated system are soil moisture sensors, a motor to pump water and a NodeMCU platform. Advancements in smart farming in collaboration with fog computing have also shifted the trend from traditional to new advanced techniques for farming. Precision agriculture is one of the methods used alongside smart irrigation and other developed farming techniques. Various tools and technologies used for precision agriculture or smart farming include sensing technologies such as soil sensors, light, water scanning, and temperature and humidity sensors, software technologies that provide solutions such as cellular communication, NFC (near field communication), positioning technology like GPS tracking system, hardware and software enabling IoT-based solutions at the same time sharing data to and from the cloud, data mining and analytics for gathering relevant data and then making decisions based on them also using them for prediction analysis, future trends analysis and historical analysis [10]. Equipped with all possible tools, farmers can monitor and regulate field conditions without going to the field. All these implementations of smart farming techniques and automating devices help to ease the burden on the farmer. The IoT-based farming cycle consists of four major processes, namely observation, diagnostics, decisions and implementation [9]. The first step involves recording the sensor data by observing the environmental surroundings such as the soil, air, water, light and crops. The second step involves analysis of the collected data, which are fed to specific software and processed with certain predefined rules and algorithms. After the diagnostics of the data, issues are revealed that are dealt with by either location-specific treatments or solutions or by general solutions. The last step is the implementation of the solutions, which need to be automated in the case of IoT so they depend on the machinery used and the communication and connection between different machines.

Another advancement is automation in smart greenhouses. Traditional greenhouses control environmental parameters manually or through proportional control mechanisms, which often lead to monetary losses, waste of materials such as fertilizers, productivity loss, energy loss and high maintenance costs [11], whereas the automated smart greenhouse controls the mechanism for regulating the environmental parameters such as light, temperature and climate, eliminating the need for manual intervention and decreasing the cost of labor and physical stress. To do this, different sensors collect data from various plants surrounding the greenhouse and then send the data to the cloud or the intermediate fog layer for processing. They then provide crop-specific solutions with minimal manual labor [12].

Also, drip irrigation and automated smart irrigation are something new brought to the floor due to the advancement of technology that can also be used for medicinal plants. Plants included in this category are aloe vera, rosemary, basil, thyme, jojoba and tulsi. These plants are very useful in the preparation of various medicines and drugs and also in homemade medicinal therapies. They require little water to grow; hence, this system works because it is automated and minimal water can be provided.

Agricultural drones are also a major breakthrough in the science and discovery field of farming. Both ground-based and aerial drones are being used for crop health assessment, crop monitoring, irrigation, planting, soil and field analysis, crop spraying and other specific areas [13]. Drones work by collecting a variety of data such as multispectral, visual and thermal imagery [8]. This collected data is shared to the cloud, which is then accessed by the farmers to provide them insights into plant health indices, yield prediction, scouting reports, damage such as infection to the plants by molds or fungi reporting, chlorophyll measurement, nitrogen content measurement, stockpile measuring, field water ponding mapping, canopy cover mapping, drainage mapping, weed pressure mapping, plant counting, plant height measurement, and so on and so forth [3]. This type of system can not only add value to large-scale farms but also to small-scale farms used for organic farming, family farming, growing specific cultures, hybrid cultivation, preservation of high-quality variety and making farming more efficient and effective.

"Precision agronomics" is a term used when combining methodology with technology. It involves accessing real-time data and providing more accurate solutions for planting and growing crops. The various components or elements of precision agronomics are called "variable rate technology" (VRT), which is used to allow farmers to apply variable inputs in specific locations [14]. It includes a DGPS (differential global positioning system), which enables it to be used in three different ways: map-based, sensor-based and manual. GPS soil sampling is another element of precision agronomics. It is used for testing nutrients available in particular locations, the measure of pH of that area, and a range of other data important for making informed and profitable decisions [8]. It allows farmers to consider productivity differences within a field and formulate a plan for farming taking these differences into account. There also has been development in a remote measurement and control system for greenhouses based on GSM-SMS [15]. It includes a PC-based database system that is connected to a base station. A base station is developed using a GSM module, microcontroller, sensors and actuators. The practical application involves the central system sending and receiving messages through the GSM module [16]. The IOT SMS alarm system based on SIM900A module of SIMCOM Company was designed for an automated smart greenhouse [4]. The system gathers environmental parameters such as temperature and humidity. The use of AT command enables this system to realize SMS automatic sending and receiving, insufficient balance alarm, etc. [4]. Through this system the alarm message can be sent to the user's smartphone automatically without knowledge of the user's location.

There was also a proposed system that works on solar energy through photo-voltaic cells [17]. This system does not depend on electricity and is hence more reliable. After the observation phase and the collection of the data, based on the sensed values, PIC microcontroller is used to turn the motor pump on or off. There also exists a smart wireless sensor network for monitoring parameters using Zigbee protocol [17]. The various nodes or sources send data wirelessly to a central server, which collects data, stores it and allows it to be accessed for future analysis and then displayed as needed and sent to the client's smartphone via the internet. So, based on the above-mentioned advancements, smart farming incorporating technologies such as fog computing is the future of agriculture to increase the quality as well as the quantity of crops and to make the life of farmers easier.

6.3 Proposed Work

The proposed system would work as follows. First, the soil sensor would be placed near the root part of the medicinal plants involved, such as aloe vera, rosemary, basil and thyme. These herbs require little water to grow and possess a high grade of medicinal qualities. The soil sensor would detect the moisture level already present and, according to the sensing value, feed the information to the system, which provides water to the parts lacking water. Therefore, plants that require less water can be easily irrigated through this system. The system also includes a temperature and humidity sensor to sense the humidity of the surroundings and accordingly provide the water to the area that requires it [18]. This system includes a combination of both hardware and software components. The hardware part consists of various sensors such as a soil moisture sensor (DHT11), temperature and humidity sensor, the NodeMCU ESP8266 development board and WiFi module, a motor to pump water from the tank, relay module and jumper wires. The software part consists of the Arduino IDE, which is used to program the NodeMCU board; Thing Speak Server, which is used to store and recover data from the sensors and display them graphically; and Blynk application, which is used to control the NodeMCU board and receive information from the sensors as well as notify the user about it. Advantages of this system include saving the farmer time and money; easy and convenient maintenance; and minimizing the infrastructure to store, carry and supply water to the crops [19]. This system increases the productivity and efficiency of the existing irrigation system so that farmers can make smart decisions and also protect water resources for future generations.

NodeMCU Platform. This platform is the building block of the automated irrigation system. It is an inexpensive version of the Arduino board that is an open-source software and hardware development environment with an in-built system on chip (SoC) ESP8266 WiFi module which is required for sending data and information to the Thing Speak Server and also to the Blynk application over the internet. The memory specifications for this board are a 4 MB flash memory, which is smaller in size and hence more portable. It can be accessed or programmed through the Arduino IDE, similar to the Arduino board.

Motor. This device is dispenses water from the water tank per the commands received by the controlling system in the form of signals. It converts energy into motion. Advanced motors can even employ solar energy to meet their energy requirements, making them energy efficient.

Sensors. Sensors, as the name suggests, sense the surroundings and reports information to the controlling system. They act as an interacting interface between the physical surroundings and the controlling system. They also give detailed feedback about the movements, functioning and status of other components in the controlling system. Sensors also comprehend the various outside forces acting on the plants and coordinate the other components of the system accordingly.

Various types of sensors used:

- Soil moisture sensor: These sensors are placed in the ground to sense the soil moisture from time to time. The LED gives a sign of whether the sensors are sensing or not. It then sends the data to the controlling unit.
- Temperature and humidity sensor: This sensor felicitates the system to sense the surrounding temperature and humidity and then transmit the sensed information to the control unit.

Thing Speak Server. This open-source IoT application and API stores and recovers information from things using HTTP over the internet or through LAN. It also provides functionalities similar to the MATLAB software for computations and provides an interface for graphical representation of data.

Blynk application. This application is compatible with both IOS and Android to control development boards such as Arduino, NodeMCU and Raspberry Pi over the internet.

It provides an interface to the user to get information directly from the sensors and actuators in the form of notifications.

6.3.1 Benefits of Keeping Medicinal Plants

Indoor medicinal plants are cheap and affordable when compared to drugs. They save time and energy by being easily available at home. They have 1,000 years of scientific evidence to back them up. On a larger scale, they can be a source of continuous income if one prefers to grow them on a larger scale. Regular use of them keeps one in good health.

According to one's taste and preferences, one can choose what to grow. The best part is that it reduces dependency on drugs [6].

6.4 Proposed Algorithm

The functionality of the IoT-based irrigation system is depicted in Figure 6.1. The power is supplied to the system by connecting the NodeMCU platform with the power supply. The sensing starts, soil sensors detect the soil moisture, and the temperature and humidity sensors detect the moisture in the surroundings. Both the soil sensor and the

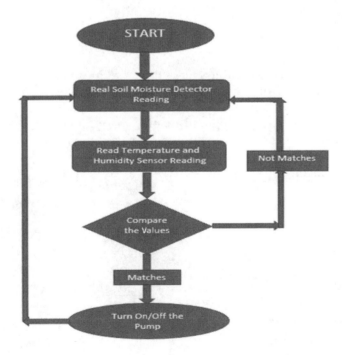

FIGURE 6.1
Flow chart of IoT-based irrigation system.

temperature and humidity sensor module send their data to the controlling unit or NodeMCU ESP8266 WiFi module [15].

Here, the comparison of values takes place with a predefined threshold value suitable for the particular crop written in the code in the Arduino IDE, which is uploaded in the NodeMCU board [3]. The soil moisture percentage is calculated and then compared with the threshold. Simultaneously, the control unit platform sends the sensed data to the Thing Speak Server and the Blynk application for further processing. The sensed value is compared to a threshold value that is feed into the system after calculations performed according to the requirements of the medicinal plants. Aloe vera is a plant that easily grows with very little water supplied to it and is highly beneficial. It is also used in all kinds of medicines, medicinal products, beauty products and hair products. Thus, this system can be used commercially in the market. Similar plants such as basil, thyme, rosemary, mint, ginger and turmeric, which require less water and have high medicinal value, can also be created through this system [20]. Only the threshold value needs to be set and the system is good to go.

- If (sensed value < threshold value), then start or turn on the water pump module [8].
- If (sensed value > = threshold value), the turn off the water pump module [8].
- If the values are unclear and the comparison fails, then the sensors re-read the values.

6.5 Results

The result is that if the sensed value is less than the threshold value, then the water pump module is turned on. If the sensed value is greater than or equal to the threshold, then the water pump module is turned off. The Thing Speak Server will store the data for future processing and display the graphical representation of the data. Figure 6.2 exhibits the soil moisture percentage graph, and Figure 6.3 displays

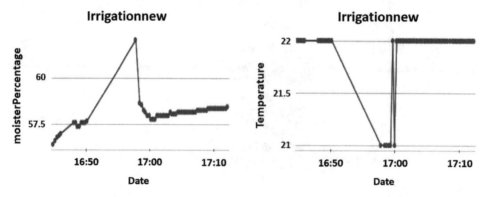

FIGURE 6.2
Soil moisture percentage graph in Thing Speak Server.

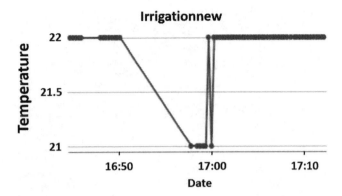

FIGURE 6.3
Temperature graph in Celsius.

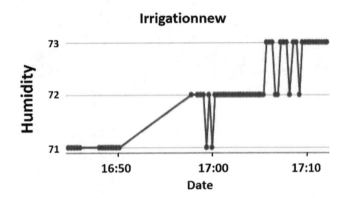

FIGURE 6.4
Humidity percentage graph.

the temperature graph in Celsius. Finally, Figure 6.4 presents the datewise humidity percentage for soil.

The Blynk application also displays (as shown in Figure 6.5) the sensor data on the user's mobile phone along with the notification of whether the water pump module is turned on or off over the internet for remote vigilance.

6.6 Conclusion and Future Scope

The proposed IoT-based irrigation system using NodeMCU is a cost-effective method for minimizing human labor and waste of water resources. The system is simple to use and is automated, which is its biggest advantage. It can be concluded that there is still scope for considerable development in irrigation with IoT and automation, and this system provides solutions to problems faced in the existing process of irrigation.

FIGURE 6.5
Blynk application soil moisture, humidity and temperature widgets.

References

1. S. B. Saraf, D. H. Gawali. (2017), "IoT based smart irrigation monitoring and controlling system," in 2nd IEEE International Conference on Recent Trends in Electronics, Information & Communication Technology (RTEICT), Bangalore, pp. 815–819.
2. M. Monica, B. Yeshika, G. S. Abhishek, et al. (2017), "IoT based control and automation of smart irrigation system: An automated irrigation system using sensors, GSM, Bluetooth, and cloud technology," in International Conference on Recent Innovations in Signal Processing and Embedded Systems (RISE), Bhopal, pp. 601–607.
3. A. Goap, D. Sharma, A. K. Shukla, C. Rama Krishna. (2018). "An IoT based smart irrigation management system using machine learning and open source technologies". *Science Direct*

[Online]. vol. 155, pp. 41–49. Available: https://www.sciencedirect.com/science/article/pii/S0168169918306987#!

4. A. Gori, M. Singh, O. Thanawala, A. Vishwakarma, A. Shaikh. (2017), "Smart irrigation system using IOT". *International Journal of Advanced Research in Computer and Communication Engineering (IJARCCE) [Online]*. vol. 6, issue 9, 1021–2278. Available: https://ijarcce.com/upload/2017/september-17/IJARCCE%2039.pdf

5. Rinkesh, (2018). "Miraculous drought tolerant plant ". *Conserve energy future [Online]*. Available: https://www.conserve-energy-future.com/drought-tolerant-plants.php

6. Anonymous. (2017). "Best medicinal plants". *Indoor plants info [Online]*. Available: https://indoorplants.info/medicinal-plants/

7. Arricca Elin Sansone. (2020). "Best healing plants". *Country Living [Online]*. Available: https://www.countryliving.com/gardening/garden-ideas/g29804807/best-healing-plants/

8. M. T. Michael Tharrington. (2015). DZone[Online]. Available: https://dzone.com/articles/the-future-of-smart-farming-with-iot-and-open-sour

9. S. Rawal. (2017). "IOT based smart irrigation system ". *International Journal of Computer Applications (0975–8887) [Online]*. vol. 159, issue 8, pp. 7–11. Available: https://www.ijcaonline.org/archives/volume159/number8/rawal-2017-ijca-913001.pdf

10. R. Nandhini, S. Poovizhi, et al. (2017), "Arduino based smart irrigation system using IOT", in 3rd National Conference on Intelligent Information and Computing Technologies, IICT '17, Coimbatore, 2017.

11. V. R. Balaji. (2019). "Smart irrigation system using IoT and image processing". *International Journal of Engineering and Advanced Technology (IJEAT) [Online]*. vol. 8, issue 6S, ISSN: 2249–8958. Available: https://www.ijeat.org/wp-content/uploads/papers/v8i6S/F10240886S19.pdf

12. A. P. Abhiemanyu Pandit. (2019). Circuit digest [Online]. Available: https://circuitdigest.com/microcontroller-projects/iot-based-smart-irrigation-system-using-esp8266-and-soil-moisture-sensor

13. B. Singla, S. Mishra, A. Singh, S. Yadav. (2018). "A study on smart irrigation system using IoT". *International Journal of Advance Research, Ideas and Innovations in Technology (IJARIIT) [Online]*. vol. 5, issue 2, ISSN: 2454-132X. Available: https://www.ijariit.com/manuscripts/v5i2/V5I2-1894.pdf

14. R. K. Yadav, R. Dev, T. Mann, T. Verma. (2018). "IoT based irrigation system". *International Journal of Engineering and Technology [Online]*. vol. 7, issue 4.5, pp. 130–133.

15. A. Jogdand, A. Chaudhari, N. Kadu, U. Shroff. (2019). "WSN based temperature monitoring system for multiple locations in the industry". *International Journal of Trend in Scientific Research and Development (IJTSRD) [Online]*. vol. 3, issue 4, e-2456–e-6470. Available: https://www.ijtsrd.com

16. G. Ravikumar, T. VenuGopal, V. Sridhar, G. Nagendra. (2018). "Smart irrigation system". *International Journal of Pure and Applied Mathematics (IJPAM) [Online]*. vol. 119, issue 15, ISSN: 1314-3395. Available: https://acadpubl.eu/hub/2018-119-15/4/724.pdf

17. P. Naik, A. Kumbi, K. Katti, N. Telkar. (2018). "Automation of irrigation system using IoT". *International Journal of Engineering and Manufacturing Science (IJEMS) [Online]*. vol. 8, issue 1, pp 77–88. Available: https://www.ripublication.com/ijems_spl/ijemsv8n1_08.pdf

18. B. R. RajeswaraRao, "Medicinal plants for dry areas". *Sustainable Alternate Land Use Systems for Drylands*, Edition: 1, Chapter: 12, pp.139–156.

19. S. Nalini Durga, M. Ramakrishna. (2018). "Smart irrigation system based on soil moisture using IOT". *International Research Journal of Engineering and Technology (IRJET) [Online]*. Volume 5, issue 6, e-ISSN: 2395-0056. Available: https://www.irjet.net/archives/V5/i6/IRJET-V5I6379.pdf

20. B. R. RajeswaraRao, R. Ramesh Kumar, et al. (2018). "Cultivation of medicinal plants". *National Seminar on Survey, Utilization and Conservation of Medicinal Plants*, pp. 2003–2007.

7

IoT Networks for Healthcare

Abhijit Paul and Medha Gupta
*Amity Institute of Information Technology
Kolkata, Amity University Kolkata, Kolkata,
India*

CONTENTS

DOI: 10.1201/9780429318078-7

7.1 Introduction

Telephone networks are used for telephone calls among two or more parties. They are primarily divided into landlines and mobile networks. In a landline network, telephone devices are directly connected through wires into the telephone exchange or public switched telephone network (PSTN). In a mobile network, mobile devices are connected wirelessly through radio frequency (RF) links with the nearest cell tower. Mobile devices can move anywhere within the coverage area. Telephone networks operate on the principle of circuit switching, but circuit switching is not efficient because resources are allocated for the entire duration of the connection and are not available to other connections. An alternative approach to circuit switching is packet switching. Advanced Research Projects Agency Network (ARPANET) uses the concept of packet switching to transmit data. ARPANET consists of a host and a subnet. The host sends messages to interface message processors (IMP); IMP breaks messages into packets and forwards them toward the destination. The role of the subnet is to store and forward packets. As more and more networks are getting connected to ARPANET, the standard model must be strengthened.

The TCP/IP model and protocol was developed to handle communication over internetwork. In TCP/IP, different networks are allowed to communicate with each other. In 1982, the internet was born. This globally connected network system uses TCP/IP to transmit data. When a device gets connected with the internet, it can send/receive data to/from another device that is already connected to the internet. Types of data that can be communicated using the internet are text, graphics, audio, video, computer program, etc. The internet provides global exchanges of data among private, public, business, academic and government networks. After the development of the World Wide Web (WWW) in 1989, the popularity of the network increased, and the number of users increased exponentially, which led to an increase in traffic (flow of data) in the network. The explosion of social media after 2005 further boosted traffic. Global internet traffic in 1990 was 0.001 petabyte (PB) per month. It became 75 PM/month in 2000. It was 2,055 PM/month in 2005, 14,955 PB/month in 2010, and 49,494 PB/month in 2015. It is expected to be 174 Exabyte (EB) per month in 2020 [1].

The desire to connect objects to communicate and share data without human interaction created the Internet of Things (IoT). IoT refers to a worldwide network of interconnected heterogeneous objects that are uniquely addressable, based on standard communication protocols [2]. Heterogeneous objects include sensors, actuators, smart devices, smart objects, radio-frequency identification (RFID) devices and embedded computers. This term was first introduced by Kevin Ashton in 1999. As per CISCO, IoT was born between 2008 and 2009 [3]. IoT-enabled objects are connected to the network and can communicate information with each other and can control an entity from a distance. This technology converted the existing internet infrastructure into a highly superior computing network, where devices are identified and connected with each other. IoT extends network connectivity beyond normal devices such as desktops, laptops, smartphones and tablets to a wide range of devices. IoT-enabled devices use built-in technology to communicate and interact with the external environment.

IoT is significant for next-generation technologies that can impact the whole business spectrum and can be thought of as the interconnection of uniquely identifiable smart objects and devices. This can lead to an advantage over the internet infrastructure. It goes beyond machine-to-machine (M2M) scenarios by implementing advanced connectivity of devices, systems and services. It also provides appropriate solutions for a wide range of applications such as smart cities, traffic congestion, waste management, structural health, security, emergency services, logistics, retail, industrial control and healthcare [4]. IoT has the potential to give rise to many medical issues such as remote health monitoring [5], fitness programs, chronic diseases treatment and elderly care. Monitoring of compliance with treatment and medication at home is another important application. Therefore, various medical devices, sensors and imaging devices can be viewed as smart devices or objects constituting the most important part of the IoT.

IoT-based healthcare services are expected to reduce costs, increase quality of life and enrich the user's experience [6]. From the perspective of healthcare providers, IoT has the potential to reduce device throughput through remote provision. In addition, IoT can correctly identify optimum times for reducing supplies for various devices for their smooth and continuous operation. Economic development, technological development and social development are also dependent on improving human health. Rapid population increase has caused great pressure on the food supply and healthcare systems all over the world. IoT is expected to offer promising solutions to various problems and issues faced in everyday life.

The IoT healthcare network or the IoT network for healthcare is one of the vital elements in the healthcare system. It supports access to the IoT backbone, encourages reception and transmission of medical data and enables healthcare-tailored communications.

7.2 Data Communication in IoT

IoT devices are typically connected to the internet through the internet protocol (IP) stack. Computing in the stack is complex because it demands a large amount of power and memory from the connected devices. IoT devices can also be connected locally through non-IP networks, where devices are connected to the internet via a smart gateway, which consumes less power. Non-IP communication channels are RFID, Bluetooth, near-field communication (NFC), etc. These channels are popular but have a low coverage area, up to a few meters only. Therefore, IoT applications are limited to small area networks. Personal area network (PAN) is widely used for IoT applications. It is used for personal devices to communicate among themselves or connect to a higher level network and the internet when one master device serves as a gateway. A few leading communication technologies used in the IoT world are RFID, NFC, IEEE 802.15.4, low-power WiFi, 6LoWPAN, Sigfox and LoraWAN [7].

7.2.1 RFID

RFID uses radio waves to uniquely and automatically identify an object. An RFID system consists of four components: a tag, a transponder, an interrogator and a reader. An RFID tag consists of a microchip, memory and an antenna. An RFID tag can be incorporated into a product for the purpose of identification using radio waves. The transponder emits a message with an identification number that is retrieved from a database and uses writable memory to transmit that message among RFID readers in different locations. The interrogator is an antenna packaged with a transceiver and decoder. It emits a signal that activates the RFID tag so that it can read and write data to it. When an RFID tag passes through the electromagnetic zone, the reader detects an activation signal. The reader decodes the data encoded in the tag circuit and passes it to the host computer. Then, application software processes the data from the host computer.

The functions of the RFID system generally include three aspects: monitoring, tracking and supervising. Monitoring refers to repeatedly observing the particular conditions, especially to detect them and give warning of change. Tracking refers to observing a person or object on the move and supplying respective location data. Supervising is monitoring the behaviors, activities or other changing information, usually of people. It is sometimes done in a secret or in inconspicuous manner. The RFID applications are numerous and far reaching. The most interesting and successful applications include supply chain management, production process control and object tracking management.

IoT refers to uniquely identifiable objects (or things) and their virtual representations in an internet-like structure. The IoT first became popular through the Auto-ID Center and related market analyst publications. RFID is often seen as a prerequisite for the IoT. If all objects of daily life were equipped with radio tags, they could be identified and inventoried by computers.

7.2.2 Near-Field Communication (NFC)

NFC is a wireless technology that allows a device to collect and interpret data from another closely located NFC device or tag. This standard specifies a way for the devices to establish a peer-to-peer (P2P) network to exchange data. NFC is often used by mobile phones to read information when they pass very close to another NFC device or NFC tag. NFC uses electromagnetic induction between two loop antennas when information exchange takes place between the NFC-enabled devices. It operates within a globally available unlicensed radio frequency ISM band of 13.56 MHz on ISO/IEC 18000-3 air interface at rates ranging from 106kbit/s to 424 kbit/s [8].

As wide-open network has higher chance to get hacked, NFC counters such provision with built-in features that limit opportunities for eavesdropping. It also has easy-todeploy options for additional protections to match each use case. NFC technology has the capacity to solve the problem of unpowered objects that lack network access. For unpowered and unconnected objects, it can open a uniform resource locator (URL) and provide access to online information through a tap of an NFC-enabled device.

NFC is playing a key role in making the IoT a working reality with 38.5 billion connected devices expected by 2020 and over 1 billion NFC-enabled devices already on the market.

7.2.3 Bluetooth

Bluetooth is a wireless technology standard for exchanging data between fixed and mobile devices over short distances. It uses short-wavelength ultra high frequency (UHF) radio waves in the industrial, scientific and medical fields, but it uses 2.400–2.485 GHz radio bands for building PANs.

There are two categories of Bluetooth devices: Bluetooth classic, Bluetooth low energy (BLE). BLE is more prominent in applications where power consumption is a major issue such as battery-powered devices and where small amounts of data are transferred infrequently such as in sensor applications. Unfortunately, both categories of Bluetooth devices are incompatible with each other. Therefore, to ensure communication between them, devices must support dual categories.

In many cases, IoT technology deals with battery-powered smaller devices such as sensors, so power consumption is a crucial issue. Because BLE has many features that favor minimum power consumption, it is a more common and effective protocol for IoT. Bluetooth version 5.0 introduced many features considering low power consumption, which is useful for many IoT applications. A combination of potential new IoT applications, together with features of Bluetooth version-5, made its wide adoption inevitable.

7.2.4 WiFi

Wireless Fidelity (WiFi) is a wireless technology standard that allows user appliances to connect to the internet or stream media wirelessly. Bluetooth and WiFi both are wireless communication protocols and even share some frequencies under the ISM band. However, they differ in many parameters. WiFi has a maximum 100 m operating range, whereas Bluetooth's limit is only 10 m. Bluetooth provides easy pairing between devices and consumes less power, but it lags behind for security issues and data exchange speeds. On the other hand, WiFi has higher security and provides higher data exchange speeds, though setups are complicated and consumption of power is higher.

Though basic WiFi standards such as 802.11 a/b/g/n/ac are not the best option for IoT, few IoT applications are useful for building and campus environments. Considering such issues, another WiFi standard named WiFi HaLow was designed specifically for IoT. WiFi HaLow is based on 802.11 ah. Features that made it more suitable for IoT are lower energy consumption and ability to create of large groups of stations or sensors that cooperate to share signals. WiFi vendors are continuously working on new versions to bring more features into IoT.

7.2.5 Global System for Mobile Communication (GSM)

Global system for mobile communication (GSM) is a digital cellular technology used for transmitting mobile voice and data services, including roaming service. Roaming is the ability to use a GSM phone number in another GSM network. Data rates vary from 64 kbps to 120 mbps.

IoT devices need to be connected to mobile networks in real time and use tens of gigabytes data per month. On the other hand, GSM is not precisely optimized to handle such a large amount of devices per a small area. To deal with this problem, extended coverage GSM IoT (EC-GSM-IoT) was designed, which uses low-power, wide-area technology. EC-GSM-IoT uses a licensed spectrum and is based on enhanced general packet radio service (EGPRS). It is designed as a high capacity, long range, low energy and low complexity cellular system to support IoT. EC-GSM-IoT can be deployed via a software upgrade to existing GSM networks, thereby ensuring widespread coverage and accelerated time to market. EC-GSM-IoT-connected devices can have a battery life of up to 10 years across wide-range use cases. EC-GSM-IoT networks can co-exist with conventional cellular networks. They provide better security and privacy such as support for user identity confidentiality, entity authentication, data integrity and mobile equipment identification.

7.3 Power Conservation in IoT Devices

Relatively simple and small devices powered by batteries are required to serve for long periods of time, which extends to a few years in some cases [9,10]. To ensure long-time functioning of these small and simple devices, a proper power conservation strategy is required. To conserve power for these devices, system energy usage patterns need to be redesigned [11,12]. Design considerations include microcontroller (MCU) design, sensor operating principles, radiofrequency selection, etc. (Figure 7.1).

7.3.1 MCU Design

To make MCU energy efficient, computational requirements are given preference. Selection to arrange 32-bit or 8-bit takes place regardless of the MCU choice. In a sleep

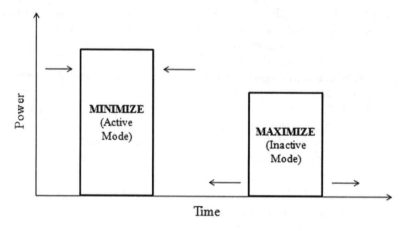

FIGURE 7.1
Optimum strategy to minimize active time.

state, energy consumption is less, but when it is changed to an active state, energy consumption is higher. Consumption of energy is comparatively less when it stays in some particular state, but it goes higher during state change. If frequency of state change is lesser, overall less energy will be consumed. MCU design and operation is mostly dependent on CPU. Numerous changes at the CPU level only to support MCU operations are not reasonable always, but significant power savings can be achieved through autonomous handling of sensor interfaces and other peripheral functions.

7.3.2 Sensor Operating Principles

Sensors read data and send it to the MCU. When data are sent without any interpretation by the sensors, MCU receives a huge amount of useless data. Energy is wated on these useless data. To conserve this unnecessary energy consumption, data are interpreted at the sensor level, and only useful data are sent to the MCU, which maximizes the battery life of the system. Also, network topology and its associated protocols are used to maintain the wireless link, and that has an impact on power consumption. In a few cases, a simple point-to-point link consumes the least power from the battery, though configuration matters.

7.3.3 Radio Frequency Selection

To develop power-efficient connected products, most of the manufacturers choose to build a radio based on the needs of the application. Bluetooth BLE radios work well for small, low-powered devices, but they need a gateway to communicate with the internet. WiFi devices require a WiFi network and router to communicate. Cellular devices are freed from the constraints of WiFi, enabling much more remote applications. When moving from Bluetooth to cellular, power required to send and receive messages increases accordingly.

7.3.4 Bluetooth BLE

Bluetooth BLE is one of the most promising low-power consumption wireless technologies for IoT applications. From physical design to model design, power consumptions are kept at a minimum level. To reduce power consumption further, BLE devices are kept in sleep mode most of the time. When an event occurs, devices wake up and transfer a short message to a gateway, PC or smartphone. This way, active power consumption is reduced to a tenth of the energy consumption of classic Bluetooth. In low-duty cycle applications, a button cell battery can provide up to a year of power.

7.3.5 WiFi

WiFi power consumption is associated directly with the amount of data the radio is transmitting. High bandwidth applications, such as downloading web pages or streaming music, consumes much more power than low bandwidth applications, such as transmitting sensor data. For high bandwidth applications, continuous power supply is required.

To conserve power for an IoT device with its wireless radio, devices need to be fully powered when used actively. Another approach to reduce power consumption is to apply "power saving mode". This mode sets the oscillator to a lower frequency to conserve power. It can be used when the device is not actively transmitting.

TABLE 7.1

Key Features and Benefits of Leading RF Technologies Used in IoT Applications

	Zigbee	**Sub-GHz**	**WiFi**	**Bluetooth Smart**
Application Focus	Monitoring and control	Monitoring and control	Web, e-mail, video	Sensors
Type of Battery	Coin cell	Coin cell	Rechargeable (Li-ION)	Coin cell
Number of Nodes	< 10 to 1000+	< 10 to 100+	< 10 to 250	< 10
Required Throughput (kbps)	< 250	< 250	< 500	< 250
Typical Range (meters)	1–100	1–1,000+	1–100	1–70
Network Topology	Self-healing mesh, star, point-to-point	Star, point-to-point	Star, point-to-point	Star, point-to-point
Optimized For	Scalability, low power, low cost	Long range, low power, low cost	Ubiquity, high throughput	Ubiquity, low power, low cost

7.3.6 Cellular

Cellular technologies [13] such as 2G/3G/LTE mirror WiFi's power consumption strategy by reducing frequency of data usage. To reduce power consumption in a cellular network, it is important to choose an efficient and secure communication protocol that requires minimum overhead. Implementing dedicated application program interface (API) at the endpoints in the cloud can reduce data usage [14]. Also, instead of programming an IoT device to send and receive messages directly to and from open internet services, it is preferable to send smaller amounts of data using a byte-efficient communication protocol designed for IoT to the cloud, where the data can be stored, processed and retransmitted if necessary.

There are a few standard technologies, such as Zigbee, Sub-GHz, WiFi and Bluetooth Smart, that are explicitly designed and developed to ensure low power consumption, and for this specific reason they are suitable for IoT applications (Table 7.1).

Regardless of the type of connection behind an IoT product, minimizing power consumption is a big challenge. However, it is also critical to keep energy and costs of components under control. RFs are key components to conserve energy resources for IoT products. A combination of smart design and proper component selection can help developers to create cost-effective IoT products that last longer.

7.4 Technology for IoT

IoT technology refers to range of technologies, standards and applications that lead from simple connection of objects to the internet to the most complex applications [15] (Table 7.2).

IoT technologies can be categorized based on infrastructure, identification, communication, etc. (Table 7.3).

TABLE 7.2

IoT Technologies with Their Features

Technology	Frequency	Data Rate	Range	Power Usage	Cost
2G/3G	Cellular bands	10 mbps	Several Miles	High	High
Bluetooth/BLE	2.4 GHz	1, 2, 3 mbps	~300 feet	Low	Low
IEEE 802.15.4	subGHz, 2.4 GHz	40, 250 kbps	>100 square miles	Low	Low
LoRa	subGHz	<50 kbps	1–3 miles	Low	Medium
LTE Cat 0/1	Cellular Bands	1–10 mbps	Several miles	Medium	High
NB-IoT	Cellular Bands	0.1–1 mbps	Several miles	Low	High
SigFox	subGHz	<1 kbps	Several miles	Low	Medium
Weightless	subGHz	0.1–24 mbps	Several miles	Medium	Low
WiFi	subGHz, 2.4 GHz, 5 GHz	0.1–54 mbps	<300 feet	Medium	Low
Wireless HART	2.4 GHz	250 kbps	~300 feet	Medium	Medium
Zigbee	2.4 GHz	250 kbps	~300 feet	Low	Medium
Z-Wave	subGHz	40 kbps	~100 feet	Low	Medium

TABLE 7.3

IoT Technology Categories

Categories of IoT Technology	Example
Infrastructure	6LowPAN, IPv4/IPv6, RPL
Identification	EPC, uCode, IPv6, URIs
Communication	Wifi, Bluetooth, LPWAN
Discovery	Physical Web, mDNS, DNS-SD
Data protocols	MQTT, CoAP, AMQP, Websocket
Device management	TR-069, OMA-DM

7.4.1 Infrastructure

IoT infrastructure design must support various applications such as monitoring and controlling operations [16]. It also needs to increase the chances of successful transition for more and more companies to adopt IoT. Developing industry-specific IoT infrastructure can increase the possibility of many companies including their compelling projects. IPv6 over low power wireless personal area networks (6LoWPAN) is one example that provides a good infrastructure solution for implementing low-power wireless communications for IoT. It provides wireless internet connectivity at low data rates and with a low duty cycle. It can carry packet data in the form of IPv6 over IEEE 802.15.4 and other networks. It also provides end-to-end IPv6 and is able to provide direct connectivity to a huge variety of networks, including direct connectivity to the internet. It operates in the 2.4 GHz frequency range with a 250 kbps transfer rate.

7.4.2 Identification

IoT facilitates interactions between a thing and a user or between two things by electronic means. Things or a user have to be identified in order to establish interaction.

The identification process includes identification of things, identification of communication, identification of service, identification of data, identification of location. etc. Many identifier standards already exist, and many of them are applicable for specific domains, sets of domains and usage scenarios. Vehicle identification number (VIN), ISO 3779, is a thing identifier standard that can specify a uniform identification numbering system for road vehicles. One M2M TS-0001 is an application and service identifier standard. It defines various identifiers such as identifiers for applications, application entities and common service entities. IPv6 is a communication identifier standard. IPv6 is an internet layer protocol for packet-switched internetworking and provides end-to-end datagram transmission across multiple IP networks. IETF RFC 4291 defines the addressing architecture for IPv6. IPv6 addresses use of 128-bit identifiers for interfaces (unicast) and sets of interfaces (anycast and multicast). E-mail address is a user identifier standard. IETF RFC 5322 defines the format of an internet e-mail address that consists of a string followed by the at-sign character (@) and then followed by an internet domain. Identification has a much wider scope and is relevant for many applications and entities in IoT. Various identification schemes already exist but need to be standardized and deployed in the market.

7.4.3 Communication

A communication channel enables an IoT system to communicate by combining IoT-specific protocols and methods. As discussed earlier, many communication channels are available for communication among IoT devices. A few popular and efficient communication channels are RFID, Bluetooth and NFC. Though these channels have comparatively low coverage areas, they facilitate less power consumption.

7.4.4 Discovery

IoT has different data communication capabilities in terms of protocol, hardware specification, reliability, data rates, etc. It also has diversity in its types and format of data. The diversity in things and the data produced by them impose significant challenges in connecting various IoT devices. IoT also is a major source of big data, driven by its velocity, variety, value and volume. To deal with large volumes of distributed and heterogeneous data, IoT devices need to maintain interoperability among them. Therefore, devices require an efficient discovery mechanism to discover available resources. The discovery process facilitates applications to find and access data without referring to the actual source of data or location. Multicast domain name system (mDNS) is used to discover host names. mDNS resolves host names to IP addresses within small networks without a local name server. Universal plug and play (UPnP) can be used to discover each other's presence on the network. UPnP can also establish functional network services for data sharing, communications and entertainment.

7.4.5 Data Protocols

Many data protocols are available to connect various IoT devices such as message queuing telemetry transport (MQTT), advanced message queuing protocol (AMQP), etc. MQTT is specified for the supervisory control and data acquisition (SCADA) system and uses the publish or subscribe mechanism to minimize the payload and overhead with application-specific or binary formats. AMQP can authenticate a vast number of messaging applications and communication designs. It provides flow-controlled, message-oriented communication with built-in options for message delivery guarantees.

7.4.6 Device Management

An increase in trend in IoT devices, IoT's changing architecture and several approaches for data management drive proper management of IoT devices. IoT device management includes device provisioning, administration, monitoring and diagnostics. IoT device management can be handled in many ways. IoT device lifecycle management is one of the solutions where customer requirements and the essential security challenges are tackled with proper management methods. Standardized device management protocols that can be applied to IoT devices are TR-069, OMA DM, LWM2M, etc. TR-069 is a standard device management protocol based on simple object access protocol (SOAP) for managing broadband equipment, including modems, routers, gateways and home devices.

7.5 IoT Device Integration Protocol and Middleware

Complex integration [17] of different protocols and standards makes IoT applications unavailable to the general public. Even though IoT devices such as sensors and smart transducers are relatively cheap, the cost to build an IoT application is higher. Reducing the cost and complexity is a massive challenge that needs to be solved. Interoperability of heterogeneous IoT devices needs well-defined standards. The solution is to have a proper middleware platform that will hide the details of the smart things and act as a software bridge between the things and the applications [18]. Middleware abstracts hardware complexity and provides an API for communication, data management, computation, security and privacy.

IoT middleware has many superior features that make communication between the devices flexible and easier. A few prime features are interoperability, device discovery, device management, scalability, big data analytics, security and privacy. With the help of middleware services, interoperability made collaboration and information exchange easy between heterogeneous devices. Considering the dynamic nature of IoT infrastructure, device management provides solutions based on their levels of battery power and reports problems in devices to the users. Through scalability, large number of devices can communicate in an IoT setup. Ever-increasing requirements of IoT devices and applications need to be scaled. In big data analytics, big data algorithms are used to analyze huge amounts of data collected by sensors. In many cases, data related to personal life is received in such IoT applications. Middleware is capable of addressing security and privacy issues through its built-in mechanism. Middleware also has a built-in mechanism for user authentication and access control.

There are many standard middleware solutions available for the IoT that address many issues. A few popular IoT middleware solutions are FiWare, OpenIoT, etc. FiWare is a very popular IoT middleware framework that is promoted by the European Union. It contains a large body of code, reusable modules, and APIs contributed by thousands of FiWare developers. FiWare provides APIs to store the context and to query it. It supplies context to systems using context adapters based on the requirements of the destination nodes. It can also control the behavior of the IoT devices and can configure them.

OpenIoT is another popular open source solution. It can collect data from IoT devices and preprocess them. It has different APIs to interface with different types of physical nodes and get information from them.

7.6 Reliable and Secure Communication

Reliable communication refers to the type of communication where messages are guaranteed to reach their destination uncorrupted and exactly the same way they were was sent [19]. Data delivery is reliably done through a combination of sequence numbers and acknowledgment messages. If any incident such as collision or failure of data packet happens during the process, the source can recover the problem by re-transmitting the packet without any user intervention. This mode of reliable communication guarantees the delivery of a message only once. This mode is normally used when one has to depend on the underlying transport to guarantee message delivery to its destination.

Likewise, secure communication ensures safe data communication, such as secure electronic transaction with varying degrees of certainty that third parties cannot intercept what was committed [20–22]. The reliability and security mechanism is built on top of an ordinary protocol by adding sequence quantity and checksum to check every packet prior to the transmission or after reception of the packet. For a secure connection, the two most relevant security procedures are used, which are discussed below [23].

7.6.1 Authentication at Network Entry

IoT devices such as remote hosts or remote sensors can authenticate a remote server through a neighbor [24]. The authentication process occurs between the IoT device and a remote authentication server. As a result of successful authentication, a master session key (MSK) is established between the IoT device and the authentication server and then is carried from the authentication server to the gateway. A PaC-EP master key (PEMK) secret is then derived from this MSK at the gateway and delivered to the network access control enforcement point (NACEP). PEMK bootstraps the security association between the IoT device and the NACEP (Figure 7.2).

7.6.2 Secured Connection to a Distant Peer

Secured connection to a distant peer is maintained for the establishment of a secure channel by an IoT node to an endpoint outside the signified domain.

Security association such as IPsec, SSL/TLS, DTLS, etc., can be used from the gateway to the unconstrained node to perform the needed protocol translation by decoding the data coming from one segment and re-encoding it into another segment. This allows one to tailor the security level to each segment. Another option consists of having the initiating IoT node agree on a shared secret key mechanism with the remote (unconstrained) peer. This procedure requires the preliminary establishment of a secure channel from the IoT node and the gateway. In this case, once the shared secret is established, the IoT node communicates directly with the remote peer using a single end-to-end secure channel.

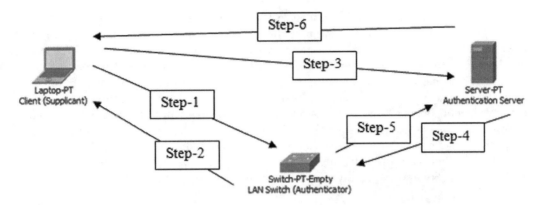

FIGURE 7.2
Flow diagram of the network entry procedure [Step-1: Request authentication, Step-2: Pass recommendation to clients, Step-3: Request service using recommendation, Step-4: Request verification of recommendation, Step-5: Verification of recommendation, Step-6: Provide service].

7.7 IoT in Healthcare

Use of an advanced technology-based healthcare method can lead to improved quality and efficiency of treatments and accordingly improve the health of patients [25–27]. IoT comes under this advanced technology and has opened many possibilities and opportunities in the field of healthcare. In the next decade, there is a higher chance of using more IoT devices for treatment and diagnosis of various diseases. In many cases, ordinary medical devices collect unintentional data, and continuous monitoring of patient status is not feasible. Such limitations are addressed in IoT technology where devices offer extra insight into symptoms, precautions and trends, and enable remote care. Though at the end of the day, medical centers need to function more competently for better treatment, implementation of IoT can make it easier and also can improve such services. The examples below demonstrate how treatment procedure will advance in the coming years.

7.7.1 Cancer Treatment

Cancer treatments can be enhanced by attaching WSNs to patients so that health practitioners can be alerted of any changes, complications, adverse drug effects and allergies, missed medications, hemoglobin-level issues, drug allergic reactions, drug interactions, etc.

The wireless sensors detect automated alerts and block incorrect prescriptions. They also automatically monitor and manage the creatinine value used in the computation of the glomerular filtration rate (GFR) used in dosing certain chemotherapy drugs. Cancer happens when cells do not have normal growth and spread very fast. Infected cells continue to grow and divide out of control, and they do not die when they are supposed to. Sensor nodes using IoT technology can detect such irregular changes or abnormal growth in the body. These sensor nodes send information to the server, which helps in early detection of diseases like cancer. In June 2018, data was presented at the American Society of Clinical Oncology (ASCO) annual meeting from a randomized clinical trial of 357 patients receiving treatment for head and neck cancer. The trial used a Bluetooth-enabled weight scale and blood pressure cuff, which was used together with a symptom-tracking app, to send updates to the patients' physicians on the received symptoms and their responses to the required treatment every day except the weekends. CYCORE is another smart monitoring system for cancer patients where less severe symptoms related to both the cancer and its treatment are compared to a control group of patients who carried on with their regular weekly physician visits. Use of smart technology improves communication between the patients and their physicians, and also monitors patients' conditions in a way that causes minimal interference with their daily lives. Bruce E. Johnson, President of ASCO, said that smart technology helped simplify care for both patients and their care providers by enabling emerging side effects to be identified and addressed quickly and efficiently to ease the burden of treatment.

7.7.2 CGM and InPen

In diabetes, the body's ability to produce or respond to the hormone insulin is impaired. Multiple risk factors make it necessary to develop new efficient strategies and appropriate prevention measures for diabetes control. Smart devices and the implementation of IoT in healthcare serve the purpose of continual monitoring and administration of treatment. A continuous glucose monitoring device (CGM) is a device that helps patients with diabetes to monitor their blood glucose level at regular intervals throughout the day by automatically tracking the individual's blood glucose level. It works through a tiny sensor that is inserted underneath the skin. The sensor tests the glucose level every few minutes. There is a transmitter that sends the required information to the monitor. Smart CGMs, such as Eversense and Freestyle Libre, send data on monitoring the blood glucose levels of the patients to an app on iPhone, Android or Apple Watch, allowing the wearer to easily check his or her information and detect trends.

Another smart device that is helping patients with diabetes is the InsulinPen (InPen). It tracks each insulin dose and delivers the data to a secure app on the smartphone. The InPen displays the active insulin blood glucose level. It also provides reminders, recommends an individual's next dose and records the amount and type of insulin injected. These devices interact with a smartphone app that can store long-term data, help patients calculate their insulin dose, and even allow patients to record their meals and blood sugar levels to see how their food and insulin intake are affecting their blood sugar level.

7.7.3 Artificial Pancreas and OpenAPS

Artificial pancreas is a technology that helps patients with diabetes by continuously control their blood glucose level by providing the substitute endocrine functionality of a healthy pancreas. Artificial pancreas is the most fascinating idea in the field of IoT medicine for the treatment of diabetes mellitus.

Open artificial pancreas system (OpenAPS) is another efficient technology that can measure the amount of glucose in a patient's bloodstream, and if needed, it injects insulin into the patient's body. OpenAPS differs from continuous glucose monitoring (CGM) because it follows a closed-loop insulin delivery system. OpenAPS software can run on a small computer such as Raspberry-Pi or Intel Edison, and it automates an insulin pump's insulin delivery to keep the blood-glucose level in a target range. The use of OpenAPS improves glycemic control while reducing the risk of grave fluctuations in the blood-glucose level of adults, adolescents and children.

Different algorithms are used in OpenAPS to affect control throughout the day, thus allowing the patient to get a good sleep without the risk of his or her blood sugar dropping to an extremely low level. OpenAPS is attracting a growing community of patients with diabetes who are using its free and open-source technology to keep track of their insulin status.

7.7.4 Connected Inhalers

Asthma is a chronic lung disease that inflames and narrows the airways in the lungs [28]. Asthma affects hundreds to millions of people every year throughout the globe. Smart and advanced technology being implemented in this field gives individuals better control over their symptoms. Smart inhalers are built with extra digital features. They contain sensors to read the condition.

The biggest producer of smart inhaler technology is Propeller Health. Rather than producing entire inhalers, Propeller has created a sensor that attaches itself to an inhaler. It connects to an app and helps people with asthma to understand what might be causing the symptoms and also provides allergen forecasts.

Using connected inhalers helps patients to be consistent about taking their medication. Besides alerting the individual's smartphone about extreme environmental conditions (dense pollution), it also checks whether a patient is taking his or her inhaler on time, thus reducing risks. The Propeller sensor generates reports on inhaler use. It can check whether patients are using inhalers as often as prescribed. Connected inhalers also motivate patients by showing their recovery phase and how use of inhalers is directly improving the patients' health conditions.

7.7.4 Ingestible Sensors

In many cases, patients do not complete their full course of medicine as prescribed by a physician. IoT technology can be used to handle the situation; ingestible sensors will monitor the patient's adherence.

Proteus Digital Health is a technology solution focused on developing products, services and data systems based on integrating medicines with ingestible, wearable, mobile and cloud computing. The system primarily consists of a smartphone, a sensor patch and a pill. When swallowed, the pill activates an attached sensor and then transmits a signal to a smartphone app. Ingestible sensors track how often patients take their medication. These pills are safe. At any time, patients can discontinue sharing private information or opt out of the program altogether.

7.7.5 Connected Contact Lenses

Medical smart contact lenses are an ambitious application of IoT. A team from Purdue University, in Indiana, came up with the idea of combining contact lenses and sensor technology to help monitor the state of blood glucose.

These wireless smart-connected contact lenses have the ability to monitor the status of the wearer's health wirelessly (glucose level in tears) through the attached LED pixel. They also provides a system for patients with diabetes to warn themselves about any irregular behavior. Besides, monitoring the blood-glucose level, the soft contact lenses are a part of the management of chronic ocular diseases.

Sensimed, a Swiss-based company, has also developed a noninvasive smart contact lens called Triggerfish, which automatically records changes in eye dimensions that can lead to glaucoma.

7.7.6 The Apple Watch App that Monitors Depression

Depression is said to be one of the most fatal psychological disease that thousands of people, especially teenagers, suffer from across the globe.

Keeping the depth of depression in mind, Takeda Pharmaceuticals joined forces with the U.K.-based Cognition Kit Limited to use the Cognition app to monitor cognitive function through the user interface on the Apple Watch. Apple Watch combines biometric data with the user input to study chronic conditions such a Parkinson's disease. The Cognitive Kit app can also measure the user's memory, attention and other essentials that help them to execute functions accordingly. Like other smart medical devices that gather data, the Apple Watch app can also give patients and healthcare professionals more insight into their condition, and enable better-informed conversations that lead to improved treatment for patients.

7.7.7 Coagulation Testing

Coagulation tests such as prothombin time (PT) and activated partial thromboplastin time (aPTT) are conducted to assess blood clotting functions in patients.

In 2016, Roche launched a Bluetooth-enabled coagulation system that allows patients to check how quickly their blood clots. The device allows patients to conduct self-tests to check for the coagulated condition. Besides checking for the coagulation in their blood, through these devices, patients can also add comments to the results of the tests. Like any other smart devices, these coagulation testing devices also remind patients to test themselves and flag the results in relation to the target range.

7.7.8 Apple's Research Kit and Parkinson's Disease

Apple added a new Movement Disorder API in 2018 to its open source Research Kit framework that allows Apple watches to continuously monitor Parkinson's disease symptoms. Research Kit offers continuous monitoring via the Apple watch. The API monitors two very grave symptoms of Parkinson's: tremors and dyskinesia. Normally symptoms are monitored by a physician at a clinic via physical diagnostic tests, and patients are encouraged to keep track in a diary in order to give a broader insight into symptoms over time. The goal of the API is to make the whole process of diagnosis automatic, continuous and easy. An app on a connected iPhone can present the data in the format of a graph giving very precise information and updates about the symptoms and treatment of Parkinson's. Apple's Research Kit has also been used in other health studies such as an arthritis study and an epilepsy study that used sensors in an Apple watch to track seizures.

7.7.9 ADAMM Asthma Monitor

ADAMM is an automated device for asthma monitoring and management. ADAMM has the ability to detect possible asthma attacks by considering all the symptoms leading to an attack. It has wearable patch type with rechargeable battery. Devices are in general worn on the upper torso, front or back. Devices catch symptoms by counting the number of coughs the patient experiences throughout the day and tracking the patient's respiration and heart rates. The device vibrates to notify the wearer about an impending asthma attack. The device has a built-in mechanism to learn what is "normal" and "usual" for the wearer over time. The device can detect and track the use of an inhaler, and it has voice journaling so the user can record any changes in feelings, behaviors, etc. ADAMM works in collaboration with a smart app that helps patients to set medication reminders along with their treatment plans.

There are obvious concerns of vulnerability involved with IoT-connected healthcare. Also, harshness of drug development is slowing down the development of new digital medicines. It is obvious that from adherence to diagnosis, IoT-based applications are suitable to get better service. In particular, life logging seems to be a superior idea, where patient treatment at a clinic can be changed. Ultimately, many leading organizations will start implementing IoT in the health sector. Apple is getting into this space with Health Kit, Research Kit and CareKit, and Google is as well with Google Fit and subsidiaries like Verily. It is not hard to imagine a future in which iOS or Android apps interact with the medical field.

7.8 Challenges in the IoT Healthcare Sector

IoT has significantly changed healthcare in a relatively short period of time. IoT-enabled devices monitor patients' vital signs and help doctors to confer with specialists across the world about chronic diseases and complex cases. However, implementing advanced technology has its challenges. A few challenges for implementing IoT in healthcare [29] are mentioned below:

7.8.1 Most IoT Initiatives Are Incomplete or Unsuccessful

In spite of discussion about IoT, there are still few initiatives for implementation and integration of IoT devices into healthcare. The research by leading stakeholder companies in the United States, the United Kingdom, India [30], etc., concluded that out of all the initiatives taken, only 26% were able to yield successful outcomes, and the rest were discarded due to problems faced in the proof-of-concept stage or shortly thereafter. To overcome the problems faced as well as increase the number of successful initiatives, organizations dealing with IoT need to be cautious when planning their IoT rollouts. For example, they should start small and prioritize projects that align with their most prominent business objectives or patient needs.

7.8.2 Healthcare Will Generate a Tremendous Amount of Data

The benefits of implementing IoT in healthcare are as follows:

- It reduces emergency room waiting times.
- It can monitor patients remotely.
- It can track many assets.
- It offers proactive alerts about medical devices that may fail soon.

IoT is a network of physical devices and other items that are embedded with electronics, software, sensors and network connectivity, which enables these objects to collect and exchange data. The health industry needs to be exceptionally careful with sensitive data gathered from various wearable devices as well as smart IoT devices according to state regulations [31]. In spite of all these advantages, IoT has integrated in the domain of healthcare. The primary backlog is that the process generates huge amounts of data. If this backlog is not taken into account, the flood of data created by the smart IoT gadgets and devices used in healthcare industry could cause many problems in the near future. Organizations need to be properly equipped to handle the large amount of data generated and also need to verify the data's quality.

7.8.3 IoT Devices Increase Available Attack Surfaces

In this world of IoT infrastructure, the attack surface is the area that can get exposed by hackers. IoT composed of all smart devices interconnects with various networks and communicates information over the networks. As uses of these devices are increasing, hackers can infiltrate the system in many ways and can extract valuable information. Along with the traditional hacking risks, a Zingbox study has identified a new risk that allows the hackers to learn about how a connected medical device operates by accessing the system and reading its error logs. This information allows hackers to break into the hospital's network or make devices publish incorrect readings, influencing the patient's treatment. Reinforcing standards and normalizing protocols can reduce the patient's risk by closing spaces between the layers of an IoT system. Facilities planning to implement IoT technology must increase their awareness of existing threats and understand how to protect networks and gadgets from hackers' efforts.

7.8.4 Outdated Infrastructure Hinders the Medical Industry

The rapid growth of IoT devices is not something that can be easily controlled. Thus, new threats and challenges get introduced every time there is an upgrade in the technology. A few ways medical industries get hindered are discussed below.

- The critical infrastructure is operated with outdated, poorly secured computers.
- The government has little authority to control the infrastructure.

As advanced technology is not known to everyone, even if the healthcare centers revamp their infrastructure, it is challenging to hire and train staff with advanced and updated technology. On the other hand, the prospective candidates may also not be capable enough to tackle the old infrastructure with ease.

7.8.5 IoT Poses Many Overlooked Obstacles

The most common use of IoT technology in healthcare is patient monitoring systems. This is undoubtedly a better way to serve the patient, but healthcare organizations often overlook the inability of smart devices to continuously go through planned periods of tasks. In addition, hospitals often depend on IoT-enabled supply to track resources. Therefore, once those systems are in place, the facilities can often reduce previous supply management issues. However, there is a chance of human error even by the smartest connected device. Many hospitals have better cybersecurity policies, but patients still may fall under risk from products that lack adequate security. One bad Apple or Android application can damage the system. Cybersecurity has to be uncompromising and complete. Unfortunately, few current IoT systems are secured by traditional network security metrics.

Although there are numerous IoT-related risks associated with healthcare IoT initiatives, health care professionals should not feel discouraged about using IoT in ways that make sense for the needs, budgets and infrastructures of their organizations. Identifying the challenges is the first step to developing effective solutions. With a deeper understanding of the challenges faced by healthcare IoT initiatives, stakeholders will be on the way to developing and deploying well-equipped healthcare IoT solutions.

7.9 Future Trends of IoT in Healthcare

Hospitals with different apps available to support them are no longer waiting for the implementation of next IoT products; rather they are in the implementation or post-implementation stages. Aruba Networks predicted 87% of healthcare organizations will be using IoT technology, 73% of applications of IoT in healthcare will be used for remote patient monitoring and maintenance, 50% for remote operation and control, and 47% for location-based services.

Many wearable devices are now being worn by patients. These devices can transmit data to physicians and thereby allowing doctors to monitor vital signs in real time, including their heart rate, glucose levels and even fall detection. These medical remote patient monitoring devices can collect key real-time data elements. Therefore the benefit is that patients will be able to in the comfort of their own homes while still under the observation by healthcare professionals.

Smart home medication dispensers have evolved which is not only notify healthcare professionals when medicine is not being taken as prescribed, but now these devices can store medicine at proper temperatures to ensure practicability.

Smart beds are making inroads in hospitals. It is collecting detailed information about patients' positions and vitals, and also helping nurses for better care for patients.

As IoT becomes more pervasive in healthcare facilities, next-generation IoT devices will bring intelligent services as part of their offering, will allow real-time data processing at the edge level or at the device level and will enable some actions to be executed by the devices if necessary, and then send back data to the patient and their clinical teams. The increasing interaction of AI and IoT in the healthcare sector is likely to move toward more intelligent IoT devices that can perform activities autonomously. This could include medical devices that react to trigger or recognize the patient and interact with them based on their treatment plan, or even the use of autonomous drones to deliver organs or drugs to facilities in need.

7.10 Conclusion

IoT has numerous applications in every field of healthcare, where various serious and deadly diseases are now handled with the utmost care. Diseases that used to take the lives of billions of people across the globe are now being treated through integration of IoT medicine. IoT has introduced wearable sensors that regularly monitor the internal environment of patients. IoT helps to improve the experience of patients by providing timely diagnosis, improved accuracy and better treatment outcomes. Ingestible sensors can verify whether the patients are taking their daily doses of medicines on time, as prescribed to them. Through automated workflow process of data transmission, error rates drop considerably compared to manual collection and reporting of data. Apart from these, IoT can also reduce the frequency of patient visits to the clinic for checkups. Integration of technology with medicine has not only made life easier but also made it worthwhile at the same time. Despite the many benefits of IoT, it has few drawbacks. Security and privacy are not guaranteed, and many smart devices are quite expensive. Still, the population is moving towards including smart devices to their healthcare routine. A 2016 survey showed that the number of patients under remote monitoring using interconnected smart devices jumped by 44% to 7.1 million.

References

1. Wikipedia (2019), "Internet Traffic", https://en.wikipedia.org/wiki/Internet_traffic, online accessed on 25June2019 at 11:20 A.M.
2. M. Zhou, G. Fortino, W. Shen, J. Mitsugi, J. Jobin and R. Bhattacharyya (2016), "Guest Editorial Special Section on Advances and Applications of Internet of Things for Smart Automated Systems," *IEEE Transactions on Automation Science and Engineering*, vol. 13, no. 3, pp. 1225–1229.
3. D. Evans (2011), "The Internet of Things: How the Next Evolution of the Internet Is Changing Everything," *Cisco Internet Business Solutions Group (IBSG), White Paper*, pp. 1–11. https://www.cisco.com/c/dam/en_us/about/ac79/docs/innov/IoT_IBSG_0411FINAL.pdf
4. P. Sethi and S. R. Sarangi (2017), "Internet of Things: Architectures, Protocols, and Applications," *Journal of Electrical and Computer Engineering*, vol. 2017, Article ID 9324035, 25 pages, https://doi.org/10.1155/2017/9324035.
5. R. K. Pathinarupothi, P. Durga and E. S. Rangan (2019), "IoT-Based Smart Edge for Global Health: Remote Monitoring With Severity Detection and Alerts Transmission," *IEEE Internet of Things Journal*, vol. 6, no. 2, pp. 2449–2462.
6. S. M. R. Islam, D. Kwak, M. H. Kabir, M. Hossain and K. Kwak (2015), "The Internet of Things for Health Care: A Comprehensive Survey," *IEEE Access*, vol. 3, pp. 678–708.
7. P. Sundaravadivel, E. Kougianos, S. P. Mohanty and M. K. Ganapathiraju (2018), "Everything You Wanted to Know about Smart Health Care: Evaluating the Different Technologies and Components of the Internet of Things for Better Health," *IEEE Consumer Electronics Magazine*, vol. 7, no. 1, pp. 18–28.
8. Wikipedia (2019), "Near Field Communications", https://en.wikipedia.org/wiki/Near-field_communication, online accessed on 25June2019 at 11:40 A.M.
9. F. Wu, K. Yang and Z. Yang (2018), "Compressed Acquisition and Denoising Recovery of EMGdi Signal in WSNs and IoT," *IEEE Transactions on Industrial Informatics*, vol. 14, no. 5, pp. 2210–2219.
10. T. Wu, F. Wu, J. Redouté and M. R. Yuce (2017), "An Autonomous Wireless Body Area Network Implementation Towards IoT Connected Healthcare Applications," *IEEE Access*, vol. 5, pp. 11413–11422.

11. M. Hooshmand, D. Zordan, D. Del Testa, E. Grisan and M. Rossi (2017), "Boosting the Battery Life of Wearables for Health Monitoring Through the Compression of Biosignals," *IEEE Internet of Things Journal*, vol. 4, no. 5, pp. 1647–1662.

12. M. Hassan, W. Hu, G. Lan, A. Seneviratne, S. Khalifa and S. K. Das (2018), "Kinetic-Powered Health Wearables: Challenges and Opportunities," *Computer*, vol. 51, no. 9, pp. 64–74.

13. M. S. Hadi, A. Q. Lawey, T. E. H. El-Gorashi and J. M. H. Elmirghani (2019), "Patient-Centric Cellular Networks Optimization Using Big Data Analytics," *IEEE Access*, vol. 7, pp. 49279–49296.

14. G. Yang et al. (2018), "IoT-Based Remote Pain Monitoring System: From Device to Cloud Platform," *IEEE Journal of Biomedical and Health Informatics*, vol. 22, no. 6, pp. 1711–1719.

15. P. A. Catherwood, D. Steele, M. Little, S. Mccomb and J. Mclaughlin (2018), "A Community-Based IoT Personalized Wireless Healthcare Solution Trial," *IEEE Journal of Translational Engineering in Health and Medicine*, vol. 6, pp. 1–13.

16. L. Catarinucci et al. (2015), "An IoT-Aware Architecture for Smart Healthcare Systems," *IEEE Internet of Things Journal*, vol. 2, no. 6, pp. 515–526.

17. H. A. Khattak, H. Farman, B. Jan and I. U. Din (2019), "Toward Integrating Vehicular Clouds with IoT for Smart City Services," *IEEE Network*, vol. 33, no. 2, pp. 65–71.

18. D. Saxena and V. Raychoudhury (2019), "Design and Verification of an NDN-Based Safety-Critical Application: A Case Study With Smart Healthcare," *IEEE Transactions on Systems, Man, and Cybernetics: Systems*, vol. 49, no. 5, pp. 991–1005.

19. P. Gope and T. Hwang (2016), "BSN-Care: A Secure IoT-Based Modern Healthcare System Using Body Sensor Network," *IEEE Sensors Journal*, vol. 16, no. 5, pp. 1368–1376.

20. A. Sawand, S. Djahel, Z. Zhang and F. NaïtAbdesselam (2015), "Toward energy-efficient and trustworthy eHealth monitoring system," *China Communications*, vol. 12, no. 1, pp. 46–65.

21. K. Yeh (2016), "A Secure IoT-Based Healthcare System With Body Sensor Networks," *IEEE Access*, vol. 4, pp. 10288–10299.

22. L. Yeh, P. Chiang, Y. Tsai and J. Huang (2018), "Cloud-Based Fine-Grained Health Information Access Control Framework for LightweightIoT Devices with Dynamic Auditing andAttribute Revocation," *IEEE Transactions on Cloud Computing*, vol. 6, no. 2, pp. 532–544.

23. R. Bonetto, N. Bui, V. Lakkundi, A. Olivereau, A. Serbanati and M. Rossi (2012), "Secure communication for smart IoT objects: Protocol stacks, use cases and practical examples," *2012 IEEE International Symposium on a World of Wireless, Mobile and Multimedia Networks (WoWMoM)*, San Francisco, CA, 2012, pp. 1–7.

24. Y. Chen, W. Sun, N. Zhang, Q. Zheng, W. Lou and Y. T. Hou (2019), "Towards Efficient Fine-Grained Access Control and Trustworthy Data Processing for Remote Monitoring Services in IoT," *IEEE Transactions on Information Forensics and Security*, vol. 14, no. 7, pp. 1830–1842.

25. P. Verma and S. K. Sood (2018), "Fog Assisted-IoT Enabled Patient Health Monitoring in Smart Homes," *IEEE Internet of Things Journal*, vol. 5, no. 3, pp. 1789–1796.

26. U. Satija, B. Ramkumarand and M. SabarimalaiManikandan (2017), "Real-Time Signal Quality-Aware ECG Telemetry System for IoT-Based Health Care Monitoring," *IEEE Internet of Things Journal*, vol. 4, no. 3, pp. 815–823.

27. C. Yi and J. Cai (2019), "A Truthful Mechanism for Scheduling Delay-Constrained Wireless Transmissions in IoT-Based Healthcare Networks," *IEEE Transactions on Wireless Communications*, vol. 18, no. 2, pp. 912–925.

28. T. Banerjee and A. Sheth (2017), "IoT Quality Control for Data and Application Needs," *IEEE Intelligent Systems*, vol. 32, no. 2, pp. 68–73.

29. H. Zhang, J. Li, B. Wen, Y. Xun and J. Liu (2018), "Connecting Intelligent Things in Smart Hospitals Using NB-IoT," *IEEE Internet of Things Journal*, vol. 5, no. 3, pp. 1550–1560.

30. V. S. Shridhar (2019), "The India of Things: Tata Communications' countrywide IoT network aims to improve traffic, manufacturing, and health care," *IEEE Spectrum*, vol. 56, no. 2, pp. 42–47.

31. S. Yang et al. (2019), "IoT Structured Long-Term Wearable Social Sensing for Mental Wellbeing," *IEEE Internet of Things Journal*, vol. 6, no. 2, pp. 3652–3662.

8

Medical Internet of Things: Techniques, Practices and Applications

Kinjal Raykarmakar, Shruti Harrison, Souvik Ghata and Anirban Das
*University of Engineering and Management,
Kolkata, India*

CONTENTS

DOI: 10.1201/9780429318078-8

8.1 Introduction: Importance of MIoT

Simultaneous monitoring and report: Real-time monitoring via connected devices can be used during medical emergencies. The IoT device can collect and transfer health data: blood pressure, oxygen, and blood sugar levels, blood pressure, etc. [1]. This data can be stored in the cloud and shared with authorized persons such as doctors or family members.

End-to-end connectivity and affordability: IoT can offer next-generation healthcare facilities resulting from information exchange and data movement.

Data assortment and analysis: IoT devices can collect, report and analyze the data in real time and cut the need to store raw data [2]. This can speed up decision making and make healthcare operations less prone to errors.

Remote medical assistance: Patients and doctors can connect, no matter how far they are from one another. Drugs can be distributed according to the patient's prescription.

8.2 Challenges in MIoT

Data security and privacy: Most IoT devices lack data protocols and standards [3]. There is also significant ambiguity regarding data ownership regulation. These factors make data accessible to cybercriminals, who can steal data and use them for their own benefit (Figure 8.1).

Integration (multiple devices and protocols): Integration of multiple devices can hinder implementation as the difference in the communication protocol complicates and reduces the efficiency of data aggregation [1].

Data overload and accuracy: With the increase in the number of records, the amount of data is so tremendous that deriving insights is difficult, which causes healthcare negligence [5].

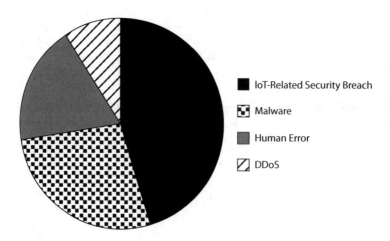

FIGURE 8.1
Challenges of IoT in healthcare [4].

8.3 Advances in MIoT

Medical Internet of Things: m-Heath familiarizes the healthcare connectivity model that connects the 6LoWPAN with evolving 4G networks [6] for future internet-based m-health services. A system for message exchange-based mobility has been introduced but its low network power consumption has not been verified.

Adverse drug reaction (ADR): ADR is damage due to drug effects. ADR is inherently generic, so an IoT framework was proposed to prevent ADR. The patient's terminal can identify the drug by means of barcode/NFC-enabled devices [7]. This information is coordinated with a pharmaceutical intelligent information system to sense whether the drug is compatible with the patient's profile.

Community healthcare: Community healthcare monitoring comes with the concept of establishing a network covering an area around a local community. This is an IoT-based network around a municipal hospital or a residential area. The structure of a community network can be viewed as a "virtual hospital" [8]. A cooperative IoT platform should be energy efficient, provide distinct authentication and use an authorization mechanism.

Wearable device access: Various non-intrusive sensors have been developed for a wide range of medical applications, particularly for WSN-based healthcare services [9]. Such sensors are prospective enough to deliver the same services through IoT.

Semantic medical access: The use of semantics and ontologies to share large amounts of medical information and knowledge can be achieved by IoT healthcare devices. Placing medial semantics and ontologies on top of IoT calls is called semantic medical access (SMA) [10].

Embedded gateway configuration: The embedded gateway configuration is an architectural service that connects patients, the internet and medical equipment [11]. The medical sensor device was developed based on IoT, and the IoT gateway is implemented through mobile computing devices.

8.4 Hardware and Software Challenges in MIoT

Hardware and software work hand in hand to make an MIoT module functional. Perfect coordination between the hardware and software is expected to help all devices work smoothly and grab precise data. However, existing challenges might pose a great threat to the proper functioning of MIoT devices.

8.4.1 Hardware Challenge

The hardware includes the physical, tangible components of the module. The hardware parts are assembled to give the module a proper shape. These include sensors, actuators and all the things that a person can interact with by touch. Some hardware challenges in MIoT are as follows:

    i. Size: A large number of sensors may require a large amount of space to accommodate them. For an MIoT device, size is a big factor. For prolonged use on a patient, an MIoT device cannot be big and heavy. All the sensors should be as compact as possible.

ii. Power management: An MIoT device needs to be powered up to work. Various sensors need electronic power to sense data. Similarly, actuators need power for their movement mechanisms. An onboard power supply must be applied with the MIoT device. Battery power might work, but MIoT health monitoring systems are typically used for longer periods of time, which would reduce the battery life and require frequent charging.

iii. Connectivity: The heart of MIoT is real-time data transfer. So, a stable, continuous connection is required. However, poor connection is always a threat due to bandwidth throttling, interference of other network peripherals, server breakdowns, etc. A dedicated channel is needed to ensure the best connection at all times. Also, different types of gateways, such as WiFi, LiFi, Bluetooth, ZigBee, IR and NFC, can be chosen based on their energy and radiation profiles.

8.4.2 Software Challenges

Hardware is a solid state, but it cannot work by itself. It requires back-end software. Bugs in software might lead to total failure of the module, even if the hardware is in working order. Some software challenges in MIoT are as follows:

i. Choosing a platform: Developing a platform from scratch would be very difficult, especially for testing and debugging. So, there are various platform options available for developers to work on. Each platform has its own pros and cons. So, it is very difficult to choose between platforms to suit all needs.

ii. Debugging: There is always a need to test the product and remove all the bugs before deploying it. To accurately test an MIoT device requires a large number of real-world test cases. Directly testing a device on a patient might be dangerous and harmful. A simulated and controlled environment can be used to test the device [12]. Maintaining such an environment might not be an easy thing. So, pseudo inputs might be helpful to test the device.

iii. Updating and compatibility: Software is continuously updated to screen for bugs, to remove security threats and to install new features. Firmware interacts with the software and hardware. Besides updating the software, users should update the firmware to ensure proper coordination between hardware and software. Sometimes, the existing hardware might not be compatible with the updated software. Users may need to change the hardware, which might lead to drastic changes in the product.

8.5 Security and Privacy Issues in MIoT

Medical devices are closely related to the human body. So, collecting and securely storing information on the patient's body is important. These devices produce a large amount of real-time data throughout the day. On the other hand, the patient's privacy information exists at all stages of data collection, data transmission, cloud storage and data republication. Undoubtedly, all MIot devices need a strong security background.

8.5.1 Data Integrity

Data transferred over a network always is at risk of being mishandled. Parts of data might get lost during transmission [13]. Also, noise in the transmission channel might lead to the transmission of incorrect information. Data integrity would be lost, leading to incorrect data. To maintain proper data integrity, an MIoT needs to validate input data, remove duplicates (based on timestamp) and audit the changes with high precision (Figure 8.2).

8.5.2 Authorization

Information about a patient's metabolism comprises sensitive information. Only authorized personnel can have access to that information. Corruption or manipulation of data by an intruder or unauthorized personnel might lead to instability of the devices, false reports or incorrect treatment for the patient. Security should be tight, with passwords, pin numbers, fingerprint authentication, etc. Also, data access levels must be set according to the assessor's degree of authority (Figure 8.3).

FIGURE 8.2
Challenge: Loss in medical data.

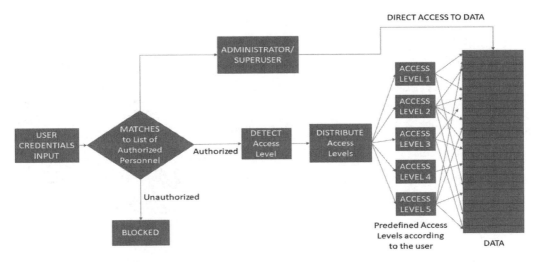

FIGURE 8.3
Flowchart showing access level of medical data.

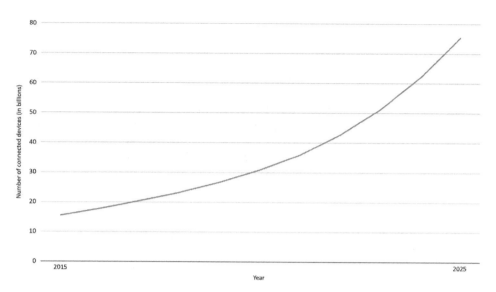

FIGURE 8.4
Number of connected devices worldwide 2015–2025 [14].

8.5.3 Cloud Storage

MIoT facilitates the use of the cloud for data transfer and storage. Issues such as breaching, insider threats, malware injection, etc., are always a point in cloud computing, and undue access to patients' sensitive information might be harmful.

8.5.4 Automation

Manually managing a large volume of data is a difficult task. Part of data handling needs to be automated using machine learning or artificial intelligence algorithms. Due to the introduction of a large number of devices, vulnerable points might increase, and it would be difficult for the same algorithms to bend to fit to new needs. This might create a security loophole.

The graph in Figure 8.4 shows the trend of IoT devices.

8.6 MIoT in Healthcare: Applications and Benefits

8.6.1 Applications

i. Wearables: Wearables are small smart devices that can be used in sync with a smartphone app that will allow real-time monitoring of the patient's health. They are technology-infused devices that can be worn on the human body. They contain sensors that facilitate the collection of raw data and relay data into a database or software [15].

ii. Moodables: These head-mounted smart devices can be worn throughout the day to enhance the mood [16]. The moodable sends out low-intensity pulses to the brain that help to elevate the mood of the wearer. These devices would be a boon for people who depend on antidepressants as they can be worn throughout the day and will not have the usual side effects that accompany the use of antidepressants.

iii. Hearables: These are the newest hearing aids that are enhancing the quality of life of people who have hearing loss. Hearables are synced with an app on the person's smartphone which allows the person to control and enhance the real-world sound with equalizers, filters and other features [17].

iv. Ingestibles: These are small capsule-sized IoT sensors that can be ingested by the patient to help the medical team monitor body irregularities and blood levels to determine if medications have been taken as prescribed. They can be particularly helpful for elderly patients.

8.6.2 Benefits

i. Improved disease management: IoT-enabled devices allow real-time monitoring of patients for things such as heart rate, blood pressure and body temperature. This facilitates the proper diagnosis of illness. If the doctor or nurse cannot be with the patient, his or her condition can be monitored remotely with an app. The IoT devices are also interconnected, which means that they can be programmed to give immediate assistance if required.

ii. Smart health monitoring: IoT-enabled smart devices allow patients to be monitored remotely. Even if the patient is not in the hospital, the smart device will collect constant data. For example, the blood pressure and temperature wearable is a small smart device that can be worn on the wrist to monitor the heartbeat, blood pressure and body temperature of the patient. The high and low indicators can be set at particular levels, which when reached will send out an alert to the app on the patient's or caregiver's smartphone and the patient can get the required medical attention right away.

iii. Connected healthcare and virtual infrastructure: The data collected by IoT-enabled devices are stored in the cloud, which makes storage and access easy for future reference or remote study. This helps the patient to receive remote healthcare, without being present at the hospital physically, hence the concept of "Virtual Hospital" [8] via IOT-enabled devices. IoT-enabled devices can go a long way toward helping people in locations where there is no proper access to hospitals and medical staff.

iv. Accurate data collection and availability: Constant automated monitoring of the patient allows for accurate data collection without the possibility of error [18]. This enhances study of the patient's condition, leading to proper diagnosis and precise treatment. IoT devices also help doctors complete patients' charts, which otherwise would take hours. With new technology such as voice commands, this can be done within hours and at an affordable price.

v. Drug management: IoT has introduced talking devices that will remind patients of the medications they need to take. This enhances the timely delivery of drugs, especially to elderly people with high blood pressure, diabetes, Alzheimer disease, etc. [7]. It also helps persons take the proper amount of the drug, which reduces deaths or problems due to wrong drug intake.

8.7 MIoT in Psychological Well-Being

The introduction of MIoT will bring a lot of changes in psychological science management. Doctors can manage patients from a distance, and through the proper implementation of MIoT, they can understand the environment of patients and treat them accordingly. Patients hopefully can make a real-world with their doctors as well.

Many technologies that are fueling IoT are also fueling neuroscience. Not just that, but IoT is directly helping neuroscience. One of the biggest promises of IoT is personalized medicine, with extensive patient records and the use of artificial intelligence helping physicians to treat future problems. IoT could one day fix the skyrocketing cost of healthcare.

Deep brain stimulation (DBS) is an operation where an electrode is implanted in the brain. The surgery has its own side effects [19], including disorientation, nausea, dizziness, sleep-related problems, and serious complications including bleeding in the brain, personality changes and diplopia. Patients with diplopia have double vision. They may think they are drunk after surgery! With long-distance control of these electrodes, in part possible with the development of technologies that are broadly lumped as IoT, doctors can make the side effects less severe. There is also another issue that IoT can fix: Sometimes implants become calcified, a problem that can be monitored and controlled with sensors that dissolve with time. Researchers at the University of Illinois at Urbana-Champaign are working on these types of implants. Another problem is that stimulators may malfunction. In 2011, Dr. P.K Doshi, a neurosurgeon, published a paper that analyzed about 153 cases of DBS that occurred from 1999 to 2009. The results were not surprising: 24 cases developed complications within 1 year, with another three cases diagnosed with a stimulator malfunction [20]. In part, the progress in DBS has been due to advances in electronics and computer science, but other aspects of IoT are also helping.

    i. Self-monitoring: Self-monitoring enables patients to control their activities by themselves and helps prevent the confusion that often happens in hospitals and clinics.

    ii. Brain-machine interface and brain mapping: The brain-machine interface is required when the IoT device is used as an interpreter. This system is mainly used for communication with a person with paralysis. A system of microelectrode arrays was implanted in a patient's brain that controlled her arm's movement so that she was able to fly an F-35 in a flight simulator.

8.8 MIoT and Cloud Computing

Cloud computing is the optimized delivery of computing services, including servers, storage, databases, networking, software, analytics and intelligence, over the internet. Cloud computing scales down operating costs and also ensures faster, smoother and more accurate operation (Figure 8.5).

Cloud computing uses the processing resources of remote servers. Servers have large storage space and high processing speeds. Therefore, small MIoT devices can rely on these robust servers for handling and processing big data without having to actually process it in their microchips. Data can be transferred from the sensors of the MIoT

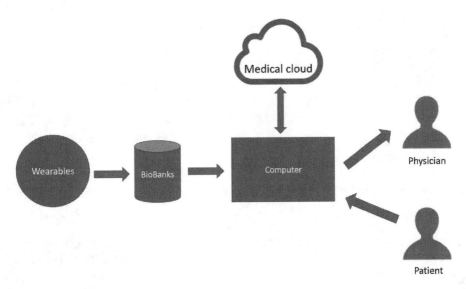

FIGURE 8.5
The MIoT cloud computing service chain.

devices directly to the server, processed and returned with generated reports. This is back-end processing.

Before pulling the data streams to the cloud, edge data stream processing is needed for detecting missing, broken and duplicated tuples in addition to recognizing tuples whose arrival time is out of order. Analytical tasks such as data filtering, data cleaning and low-level data contextualization can be executed at the edge of a network [21]. In contrast, more complex analytical tasks such as graph processing can be deployed in the cloud, and the results of ad-hoc queries and streaming graph analytics can be pushed to the edge as needed by a user application.

8.8.1 Cloud-Based Adherence Platforms Helping Patients Stay on Track

Whether the patient is taking medication at the right time, complying with dietary instructions or making it to doctor's appointments, cloud technology makes for objective reporting of real-time activity, helping providers to monitor patient behavior remotely. To help patients abide by doctor's instructions, cloud technology-based adherence platforms can help with reminders, including prompts to follow the prescribed drug regimen, which currently is one of the leading causes of avoidable readmissions [22].

8.8.2 Big Data and Predictive Technologies Enabling Preventive Action Ahead of Time

Data collected via MIoT can help identify elevated risks of developing chronic conditions at an early stage. With advanced analytics applied to big data, predictive analysis regarding the tendency and pattern of diseases across various population demographics can also be performed [23]. Representing a huge leap forward for population health management, providers, concerned administrative entities and financial experts can receive alerts about potential disease outbreaks ahead of time, helping them to make more informed choices about how to proceed with preventive measures and other life-saving activities.

8.8.3 Telemedicine Bringing Healthcare to Rural and Remote Areas

One of the biggest challenges of healthcare in India is the concentration of caregivers in urban areas, with shortages in rural and remote areas. Telemedicine is helping to address this gap by leveraging the cloud to connect patients with caregivers over smartphones and laptops, irrespective of their location. Patients in remote villages can get consultations for primary and secondary care from specialist doctors via telemedicine [22].

8.9 MIoT and Big Data

MIoT monitoring systems must gather data at all times. These data, when collected continuously and to a large extent, are no longer ordinary data, but "big data". Big data must be broken into small parts. The three Vs of big data [24] are mentioned below:

i. Volume: This defines the size of the data. Big data can be too large to handle. They might be as low as 10 terabytes, and as high as hundreds of petabytes. The size of data plays a crucial role. Before data are handled and processed, they need to be stored. Also, high processing power is required to gather this extent of data in a short time period and at regular intervals.

ii. Velocity: This stands for the data intake speed (i.e., the number of inputs taken) or data collected over a certain duration. Data are always flowing in. There is a need to collect data with a high degree of precision. MIoT collects real-time data; the velocity of big data is quite high.

iii. Variety: This refers to the different types of data that can be collected: structured, semi-structured, unstructured, numeric, character, Boolean, digital, analogue, text, audio and video. A large variety of data creates a problem for mining and analyzing. So, data need to be structured.

Big data might concern MIoT in the following points:

- MIoT provides a map of different interconnected devices and sensors. Data collected from each node dumped into a centralized point accounts for big data.

- Storing big data into packets and labeling them will not work. After collection, data needs to be analyzed. Then, queries need to be run, and intelligent decisions have to be made in real time.

- Big data is also needed to scale up, analyze trends and in turn make an accurate extrapolative predictive analysis.

- Also, unseen correlations and hidden patterns in the trends of data can be found and monitored by applying optimized sort-and-search algorithms.

- Big data will act as a large dataset for machine learning, which can be used to determine the current situation and predict the future, too. It can also compare different datasets and generate useful reports.

8.10 MIoT Innovations and Products

8.10.1 Introduction

Falls represent a major public health risk worldwide for elderly people. A fall, if not assisted in time, can cause functional impairment in an older person and a significant decrease in his or her mobility, independence and life quality. Falls and fall-induced injuries account for over 80% of all injury-related hospital admissions [25]. When a loved one falls, he or she may not be able to press a medical alert button to call for help. This is where automatic fall detection systems come in action. A fall detector is an IoT-based system available to detect when an individual has fallen and there is a need to analyze the condition and arrange immediate help for the victim. Whenever the system detects the fall of the user, it sends an SMS to the family members of the user. This ensures the person receives timely help and the situation is under control.

8.10.2 Important Modules and Sensors

8.10.2.1 MPU 6050 [26]

- Input voltage: 2.3–3.4V
- Tri-axis angular rate sensor (gyro) with a sensitivity up to 131 LSBs/dps and a full-scale range of ±250, ±500, ±1000, and ±2000dps
- Tri-axis accelerometer with a programmable full-scale range of ±2g, ±4g, ±8g, and ±16g

8.10.2.1.1 Working Principle of MPU 6050

The MPU 6050 is a sensor based on integrated three-axis MEMS (micro electro mechanical systems) technology containing an accelerometer and gyroscope. Both the accelerometer and the gyroscope are embedded inside a single chip. The MPU 6050 is a 6 DOF (degree of freedom) or a 6-axis IMU (inertial measurement unit) sensor (i.e., it will give 6 values in output—three values from the accelerometer and three from the gyroscope). This chip uses I^2C (inter-integrated circuit) protocol for communication.

- An accelerometer works on the principle of the piezoelectric effect. Whenever a person tilts the sensor the ball is supposed to move in that direction because of gravitational force. The walls are made of Piezoelectric elements, so every time the ball touches the wall an electric current will be produced, which will be interpreted in the form of values in any 3D space [27].
- A gyroscope is a spinning wheel or disc in which the axis of rotation is free to assume any orientation by itself. When rotating, the orientation of this axis is unaffected by tilting or rotation of the mounting. Because of this, gyroscopes are useful for measuring or maintaining orientation [27].

8.10.2.1.2 Flowchart

The flowchart in Figure 8.6 demonstrates the total detection algorithm for one falling instance.

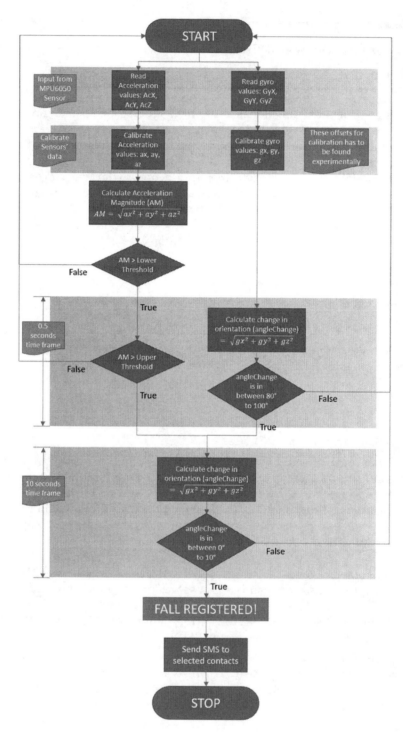

FIGURE 8.6
Flowchart demonstrating the procedure.

8.10.2.1.3 Pseudocode

```
fall_event = false
    lt_trigg = false // lower threshold breaking trigger
    ut_trigg = false // upper threshold breaking trigger
    ori_trig = false // orientation breaking trigger
    /* each trigger count accounts to 0.1 second */
    lt_trigg_count = 0
    ut_trigg_count = 0
    ori_trigg_count = 0
    AcX = AcY = AcZ = 0 // (raw data) accleration values at three direction
    ax = ay = az = 0 // calibrated values
    GyX = GyY = GyZ = 0 // (raw data) gyroscope values at three direction
    gx = gy = gz = 0 // calibrated values
    function read_mpu6050(){
    /* read the values of acceleration and gyroscopes from read_mpu6050
    and store them in (AcX, AcY, AcZ) and (GyX, GyY, GyZ) */
    }
    functioncalibrate_values(){
    /* calculateoffests via experimentation */
    ax = (AcX ± offest) / 16384.00
    ay = (AcY ± offest) / 16384.00
    az = (AcZ ± offest) / 16384.00
    gx = (GyX ± offest) / 131.07
    gy = (GyY ± offest) / 131.07
    gz = (GyZ ± offest) / 131.07
    }
    while(true){
    read_mpu6050()
    calibrate_values()
    raw_AM = pow(pow(ax,2) + pow(ay,2) + pow(az,2),0.5)
    AM = raw_AM * 10 // scalling the values from 0-10
    IFori_trig is TRUE{
    ori_trigg_count = ori_trigg_count + 1
    IFori_trigg_count> 100{
    angleChange = pow(pow(gx,2) + pow(gy,2) + pow(gz,2),0.5) // the gyroscope angle values
    IFangleChange >= 0 ANDangleChange <= 10{ // orientation change remains within 0-10 degrees
    fall_event= true// fall_event captured
    ori_trig = false
    }
    ELSE{ // Reset counter
    ori_trig = false
    ori_trigg_count = 0
    }
    }
    }
```

```
IF fall_event is TRUE{ // falling detected
send_SMS() // sends a SOS SMS to the selected contacts
}
IF ori_trigg_count>= 6{ // 0.5seconds of upper orientation changing allowed
ori_trigg = false
or_trigg_count = 0
}
IF ut_trigg_count>= 6{ // 0.5seconds of upper threshold beaking allowed
ut_trigg = false
ut_trigg_count = 0
}
IF ut_trigg is TRUE{
ut_trigg_count = ut_trigg_count + 1
angleChange = pow(pow(gx,2) + pow(gy,2) + pow(gz,2),0.5)
IF angleChange>= 30 AND angleChange<= 400{ //if orientation changes by between 80–100 degrees
ori_trigg = true
ut_trigg = false
ut_trigg_count = 0;
}
}
IF lt_trigg is TRUE{
lt_trigg_count = lt_trigg_count + 1
IF AM >= 12{ // upper threshold breaking (3g)
ut_trigg = true
lt_trigg = false
lt_trigg_count = 0
}
}
IF AM <= 2 AND ut_trigg is FALSE{ // lower threshold breaking (0.4g)
lt_trigg = true
}
pause(100) // delay of 100ms
}
function send_SMS(){
/* this function can be implemented as per the developer
other notification ways like email, app notifies etc can be set up
we have illustrated SMS here*/
read_GPS() // this function reads the GPS values. A GPS module can be
// attached to the fall detector for this case
read_local_time() // clocks can synchronised to read current dateand time
// various cloud-based IoT solutions can offer timestamp
write_SMS() // frame the text part of the SMS using collected information
FOR i in emergency_contacts{
SMS(emergency_contacts[i]) // send the SMS to the selected users
}
}
```

8.10.2.1.4 General Syntax of Notification SMS

<user_name> has fallen down at <timestamp> near <location>. Please reach the location ASAP! Follow this link to track: <link>

- <user_name>: name of the patient/user using the device
- <timestamp>: the falling time, in the format, DD/MM/YYYY HH:MM:SS
- <location>: (latitude, longitude) to pinpoint location along with Maps API driven nearby best-known location
- <link>: A Maps API handle used to live track the location of the user

8.10.2.1.5 Benefits

i. Private and affordable: Automatic detection devices provide an affordable way to ensure your parents or grandparents have emergency care just around the corner. They also give them a sense of privacy that they cannot get if they live in community care.

ii. According to statistics, the financial toll for older adult falls is expected to increase to $67.7 billion by 2020 [28]. If you want to ensure your loved ones do not have to worry about expensive medical bills caused by prolonged injuries, you need a system that will get them help as soon as needed. The automatic fall detection system is efficient, personalized and affordable, which makes it an ideal device for ensuring the safety of loved ones.

iii. Immediate help: When you use a fall detection system, you always have somebody ready to help you 24 hours a day, seven days a week. With the detection of a fall in the system, your loved one can reach out to emergency medical personnel. Users can also get immediate help from a loved ones as they will be notified as soon as the system detects a fall. This will increase peace of mind for both the user and their loved ones, knowing that they are safe and secure.

iv. Highly functional: Fall detectors are highly functional and efficient because they are portable and easy to use. In case a fall is detected or the user understands he or she is about to fall, the user can get instant access to help without having to struggle to get to the phone and type in an emergency service number or send a text to a loved ones. The fall detector is going to do it for them. As they have a GPS tracking system built within them, they can be used literally anywhere. This can be at home, at the park or even when your parents or grandparents go on vacation.

8.11 MIoT and 5G: The Future of Healthcare

Sensitive usage of IoT, such as in the medical field: We need a highly responsive network. Studying the current trends in network usage, bandwidth, and connectivity, it is clear that the current network model, 4G, needs to be improved for sustained usage of MIoT.

5G is the next generation cutting-edge network technology, promising better and reliable connections, higher bandwidth and lower latency, which is faster than the current connections. Upload/download speed up to 1 GBps is expected, which can easily handle the cloud and big data simultaneously with ease.

With low latency [29] and higher throughput communication, remote procedures such as drug releasing, surgery with haptic technology and high-definition image feedback can be done with fewer risks. Telemedical diagnosis, mobile surgery and real-time treatment would be a reality with the introduction of 5G.

5G technologies should bring hospitals, physicians and patients closer with faster link speed, thus helping them to monitor and treat better than ever. Even 5G eMBB technology (enhanced mobile broadband) can save a doctor's appointment in emergency situations [30]. This would lead doctors to care for patients all around the world without physically travelling. Also, doctors can share reports and collaborate with others with ease.

MIoT wearables, sensors, devices and apps will be connected at a faster rate. Data can be quickly acquired, communicated, processed and analyzed. Hospitals will act like data centers, with a bunch of data accessible for open study and research.

Though 5G shows a lot of prospects, it is still under development. Research on its feasibility is still on the move. With the introduction of 5G, new standards and devices will arise. The present MIoT-enabled devices may not stand a chance in competing with the new ones. Also, these old devices might lack the firmware to handle new technologies. So, they may need to be replaced with new ones:, which might be an expensive deal.

The future cannot be predicted. There might be unprecedented growth and development in the fields of engineering and medicine. These two will come together to make a better healthcare department, to treat patients well, discover more vaccines, fight chronic diseases and ultimately allow humans to live longer.

8.12 Conclusion

This chapter is introductory in nature and focuses on various aspects of MIoT. It starts with a discussion on the importance and challenges associated with MIoT-based applications. Further, the authors focus on hardware, software, security and privacy-related issues that one can encounter when working with MIoT-based system. Efficiency on computation and storing process of MIoT-based system can be improved by introducing the techniques and concepts of cloud computing and big data. The chapter concludes with a note on innovation and future healthcare issues with respect to MIoT.

References

1. P. Nasrullah, "Internet of things in healthcare: applications, benefits, and challenges," https://www.peerbits.com/, [Online]. Available: https://www.peerbits.com/blog/internet-of-things-healthcare-applications-benefits-and-challenges.html. [Accessed 17 September 2020].

2. "MEDICAL," https://dotiot.wordpress.com/, [Online]. Available: https://dotiot.wordpress.com/medical/. [Accessed 17 September 2020].

3. "IOT is coming even if the security isn't ready: here's what to do," WIRED Brand Lab for DXC Technology, [Online]. Available: https://www.wired.com/brandlab/2017/06/iot-is-coming-even-if-the-security-isnt-ready-heres-what-to-do/. [Accessed 17 September 2020].

4. S. Bhattacharjee, "Technology can move the routines of medical checks from a hospital to the patient's home — IoT in ealthcare," 18 February 2019. [Online]. Available: https://medium.com/@nite.2051993/technology-can-move-the-routines-of-medical-checks-from-a-hospital-to-the-patients-home-iot-in-95f15a972c44. [Accessed 18 September 2020].

5. D. LeSueur, "5 reasons healthcare data is unique and difficult to measure," *Explore Health Catalyst Insights*, 2014. [Online]. Available: https://www.healthcatalyst.com/insights/5-reasons-healthcare-data-is-difficult-to-measure. [Accessed 17 September 2020].

6. R. R. Bhat, 17 October 2019. [Online]. Available: https://www.a10networks.com/blog/evolution-of-iot-with-5g-future-proofing-current-iot-investment/. [Accessed 17 September 2020].

7. Z.-I. Jara, A. J. A. Skarmeta and M. A. Antoni, "Drug identification and interaction checker based on IoT to minimize adverse drug reactions and improve drug compliance," *Personal and Ubiquitous Computing*, vol. 18, pp. 5–7, 2014.

8. S. R. A. Islam, D. A. Kwak, M. H. A. Kabir, M. A. Hossain, and K. S. Kwak, "The Internet of Things for health care: a comprehensive survey," *IEEE Access*, vol. 3, pp. 678–708, 2015.

9. M. N. O. Sadiku, K. G. Ezeand S. M. Musa, "Wireless Sensor Networks for Healthcare," *Journal of Scientific and Engineering Research*, vol. 5, no. 7, pp. 210–213, 2018, ISSN: 2394-2630.

10. L. V. Bommasani, E. B. Babu, and S. Aluri, "A novel semantic medical access monitoring system for e-health applications using internet of medical things," *International Journal of Engineering & Technology*, vol. 7, p. 265, 2017.

11. S. M. R. Islam, D. Kwak, M. H. Kabir, M. Hossain and K.-S. Kwak, "The Internet of Things for health care: a comprehensive survey," *IEEE Access*, vol. 3, pp. 678–708, 2015.

12. M. Chernyak, "3 Challenges of healthcare Internet of Things (IoT) performance esting," https://hitconsultant.net/, 05 May 2019. [Online]. Available: https://hitconsultant.net/2019/05/21/3-challenges-of-healthcare-internet-of-things-iot-performance-testing/. [Accessed 17 September 2020].

13. W. Sun, Z. Cai, Y. Li, F. Liu, S. Fang and G. Wang, "Security and privacy in the medical Internet of Things: a review," *Security and Communication Networks*, vol. 2018, p. 9, 2018.

14. H. Tankovska, "Internet of Things – active connections worldwide 2015-2025," Statista, 2020. [Online]. Available: https://www.statista.com/statistics/1101442/iot-number-of-connected-devices-worldwide/. [Accessed 17 September 2020].

15. INFINITI Q50, "Smart tech that works on—and in—your body," https://mashable.com/, [Online]. Available: https://mashable.com/2015/04/30/smart-tech-on-in-body/. [Accessed 17 September 2020].

16. R. Makhija, "IoT in Healthcare," https://www.gurutechnolabs.com/, [Online]. Available: https://www.gurutechnolabs.com/iot-in-healthcare/. [Accessed 17 September 2020].

17. J. Sopher, "Are Hearables the New Wearables?," https://blog.rendia.com/, 12 January 2018. [Online]. Available: https://blog.rendia.com/hearables-new-wearables/. [Accessed 18 September 2020].

18. H. Bogdanova, "Pros and cons of remote patient onitoring," https://www.healthitoutcomes.com/, 7 February 2018. [Online]. Available: https://www.healthitoutcomes.com/doc/pros-and-cons-of-remote-patient-monitoring-0001. [Accessed 17 September 2020].

19. Mayo Clinic, "Deep brain stimulation," [Online]. Available: https://www.mayoclinic.org/tests-procedures/deep-brain-stimulation/about/pac-20384562. [Accessed 18 September 2020].

20. D. P. K. Doshi, "Expanding indications for deep brain stimulation," *Neurology India*, vol. 66, pp. 102–112, 2018.

21. I. Maduako, H. Cao, L. Hernandez and M. Wachowicz, "Combining edge and cloud computing for mobility analytics," 2017. Available at: https://arxiv.org/ftp/arxiv/papers/1706/1706.06535.pdf.

22. Sr. Director, Product Engineering at OGS, "7 ways cloud computing is shaping healthcare in 2019," ETCIO, 28 March 2019. [Online]. Available: https://cio.economictimes. indiatimes.com/news/cloud-computing/7-ways-how-cloud-computing-is-shaping-healthcare-in-2019/68614849. [Accessed 18 September 2020].
23. W. Raghupathi and V. Raghupathi, "Big data analytics in healthcare: promise and potential," *Health Information Science and Systems*, vol. 2, 2014.
24. Whishworks, "Understanding the 3 Vs of big data – volume, velocity and ariety," [Online]. Available: https://www.whishworks.com/blog/big-data/understanding-the-3-vs-of-big-data-volume-velocity-and-variety. [Accessed 17 September 2020].
25. World Health Organization, "Falls – details | fact sheets," 16 January 2018. [Online]. Available: https://www.who.int/news-room/fact-sheets/detail/falls. [Accessed 17 September 2020].
26. "MPU-6050 Six-Axis (Gyro + Accelerometer) MEMS MotionTracking™ Devices," TDK InvenSense, [Online]. Available: https://invensense.tdk.com/products/motion-tracking/6-axis/mpu-6050/. [Accessed 18 September 2020].
27. Module143, "MPU6050 – how to use it?," 18 August 2017. [Online]. Available: http://invent.module143.com/mpu6050-how-to-use-it/. [Accessed 18 September 2020].
28. National Council of Aging, "Falls prevention acts," [Online]. Available: https://www.ncoa.org/news/resources-for-reporters/get-the-facts/falls-prevention-facts/. [Accessed 17September 2020].
29. C. Gulli, "Health plan: when 5G meets IoT," Vodafone, [Online]. Available: https://www.vodafone.com.au/red-wire/health-5g-meets-iot. [Accessed 17 September 2020].
30. R. Hammer, "Top 13 innovations in healthcare echnology," 2017. [Online]. Available: https://getreferralmd.com/2018/01/future-healthcare-technology-advancements/. [Accessed 18 September 2020].

9

Issues and Aspects of Medical IoT: A Case-Based Analysis

Subrata Paul[1] and Anupam Das[2]

[1]*Research Scholar, MAKAUT, Kalyani,*
West Bengal, India
[2]*Department of CSE, ASET, Amity University*
Kolkata, Kolkata, India

CONTENTS

9.1 Introduction

The Internet of Things (IoT) "provides an integration approach for all these physical objects that contain embedded technologies to be coherently connected and enables them to communicate and sense or interact with the physical world, and also among themselves" [1]. IoT is a concept that reflects a "connected set of anyone, anything, anytime, anyplace, any service, and any network" [2].

DOI: 10.1201/9780429318078-9

IoT includes two concepts: "internet" and "thing", where "internet" refers to "The world-wide network of interconnected computer networks", based on a standard communication protocol, while "thing" refers to "an object not precisely identifiable" [3]. These concepts mean that every object can be addressable by an internet protocol (IP) and can act in a smart space such as a healthcare environment.

Another definition of IoT is "a self-configured dynamic global network infrastructure with standards and interoperable communication protocols where physical and virtual 'things' have identities, physical attributes and virtual personalities and are seamlessly integrated into the information infrastructure" [4]. Indeed, IoT is the resulting global network interconnecting smart objects by means of extended internet technologies, the set of supporting technologies necessary to realize such a vision (including, e.g., radio frequency identifications [RFIDs], sensor/actuators, machine-to-machine communication devices) and the ensemble of applications and services leveraging such technologies to open new business and market opportunities [5]. The fundamental characteristics of the IoT technology are summarized as follows:

- Real-time solutions in a global environment
- Mainly wireless solutions: indoor and outdoor environments
- Ability to remotely monitor the environment and track objects

One of the most attractive application fields for IoT is healthcare, allowing the possibility of many medical applications such as remote health monitoring, fitness programs, monitoring of chronic diseases and elderly care [2]. Healthcare is one of the main priorities for all governments. Due to population growth, rural urbanization, declining birthrate, population aging, economic growth and socially unbalanced resource utilization, social problems have become increasingly apparent in the healthcare field. Some of these issues in healthcare that IoT may prevent, or can combat in a most effective way follow:

- Health management level and the incapability of responding to an emergency
- Serious shortage in medical staff and institutional facilities, especially in rural areas; lack of medical facilities; low level of treatment; and inadequate healthcare system
- Imperfect disease prevention system that cannot meet the national strategy requirements to safeguard the health of the citizens, resulting in a heavy burden on the economy, individuals, families and the state
- Inadequate disease prevention and early detection

But there are some challenges that IoT can help to solve, such as the following:

- It breaks geographic barriers, providing rapid clinical responses.
- It provides medical consultation and communication links for medical images and video data.
- It provides a unique ontology for all things among IoT-based healthcare.

There are a lot of applications in the healthcare field, including the possibility of using smartphone capabilities as a platform for monitoring medical parameters that advise patients of medical problems.

9.2 IoT in Healthcare Networks

The IoT healthcare network or the IoT network for health care (hereafter "the IoThNet") is one of the vital elements of IoT in healthcare. It supports access to the IoT backbone, facilitates the transmission and reception of medical data and enables the use of healthcare-tailored communications.

9.2.1 The IoThNet Architecture

The IoThNet architecture refers to an outline for the specification of the IoThNet's physical elements, their functional organization and its working principles and techniques.

To start, the basic reference architecture is presented in Figure 9.1, which illustrates the telehealth and ambient assisted living systems recommended by Continua Health Alliance. The key issues have been identified for this architecture [7]: the interoperability of the IoT gateway and the wireless local area network (WLAN)/wireless personal area network (WPAN), multimedia streaming, and secure communications between IoT gateways and caregivers. Many studies [8–12] justify that the IPv6-based 6LoWPAN is the basis of the IoThNet.

As designed in Shelby and Bormann [13], Figure 9.2 shows the layer structure of the 6LoWLAN. According to the IoThNet concept, sensors and wearables use IPv6 and 6LoWPAN systems for data transmission over the 802.15.4 protocol. Data are then replied back by sensor nodes with the help of the user datagram protocol (UDP). However, the 6LoWPAN is limited in that it does not support mobile IPv6 (MIPv6), a subset of the IPv6 protocol with mobility. To introduce the mobility provision to the 6LoWPAN, a protocol for exchanging messages between mobile patient nodes, base networks and visited networks was proposed by Shahamabadi et al. [11].

To address mobility, four alternative procedures were considered in Bui et al. [14], including soliciting routers, waiting for a new directed acyclic graph (DAG)

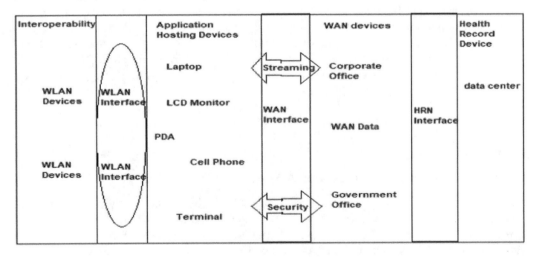

FIGURE 9.1
Continua Health Alliance's framework-based simplified reference architecture.

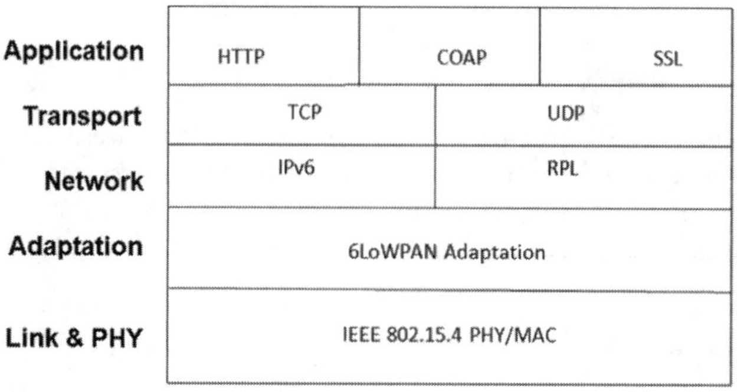

Application	HTTP		COAP		SSL
Transport	TCP			UDP	
Network	IPv6			RPL	
Adaptation	6LoWPAN Adaptation				
Link & PHY	IEEE 802.15.4 PHY/MAC				

FIGURE 9.2
The protocol stack of 6LoWPAN.

information object (DIO), attaching to other available parent nodes, and sending DAG information solicitation (DIS) messages. Among these, soliciting routers and sending DIS messages represent the fastest methods because they are initiated by the mobile node itself. A typical gateway protocol stack for community medical services is described in You et al. [15]. This stack explicitly describes how periodic traffic, abnormal traffic and query-driven traffic can be managed within the HetNet. A complex eHealth service delivery method consisting of three phases was been proposed by Swiatek and Rucinski [16], including composition, signalization and data transmission. Signalization protocols serve mainly as the basis of complex service composition, quality-of-service (QoS) negotiation and resource allocation procedures in the IoThNet. Figure 9.3 shows the state encountered in the QoS negotiation procedure, which is nothing but the creation of a connection to expected QoS values.

Medical devices have been considered for vehicular networks, and captured health data have been examined through IPv6 application servers [8]. The lightweight auto-configuration protocol shown in Figure 9.4 was introduced for vehicle-to-infrastructure (V2I) communications in the IoThNet. This protocol uses the IPv6 route as a default route in the routing table. This provides a set of IPv6 addresses for health devices in a vehicle. The extent to which big data can reshape the data structure in healthcare services is described in Diaz et al. [17], and the question of how multiple communication standards can be coordinated to give rise to the IothNet is discussed in Wang et al. [18]. The data distribution architecture is examined in the case of cloud computing integration in Xu et al. [19].

9.3 IoT Healthcare Services and Applications

Regarding IoT healthcare services and applications, the range of fields can include management of private health and fitness, care for pediatric patients, supervision of chronic diseases and monitoring of elderly patients, among others. For a better understanding of this topic, this chapter categorizes the discussion into two aspects: services and applications.

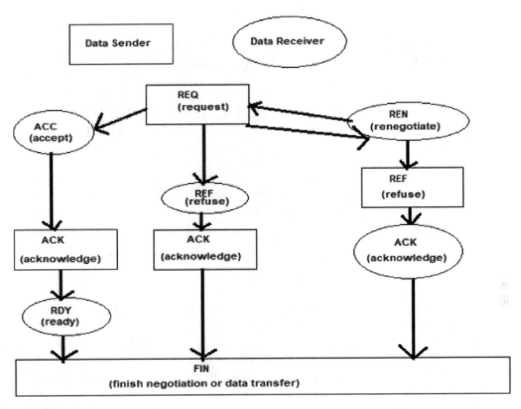

FIGURE 9.3
The negotiation process (rectangle by the sender and oval by the receiver).

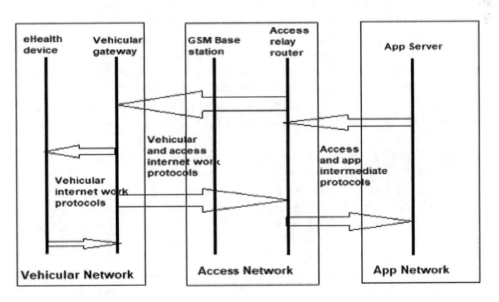

FIGURE 9.4
The auto-configuration protocol in V2I scenario.

9.3.1 Healthcare Services

i. Ambient assisted living (AAL): AAL systems have a potential to meet the personal healthcare challenges and involve citizens in their healthcare. AAL systems provide an ecosystem of medical sensors, computers, wireless networks and software applications for healthcare monitoring, a service that can be provided by IoT. That is, a separate IoT service is mandatory.

ii. m-Health things (m-IoT): The m-IoT is "defined as a new concept that matches the functionalities of m-health and IoT for a new and innovative future (4G health) applications" [20]. As shown in Istepanian et al. [21], m-health is mobile computing, medical sensors and communications technologies for healthcare services. In theory, "m-IoT familiarizes a novel healthcare connectivity model that connects the 6LoWPAN with evolving 4G networks for future internet-based m-health services. Although m-IoT characteristically represents the IoT for health care services, it is worth mentioning that there exist some specific features intrinsic to the global mobility of participating entities" [2,21].

iii. Adverse drug reaction: An adverse drug reaction is an injury from taking a medication—a single dose of a drug, its prolonged administration or as a consequence of a combination of two or more drugs [22,23]. Solutions to this issue can be found in the previous references.

iv. Community healthcare: A service that may be provided by IoT is a cooperative network structure covering an area around a local community, a municipal hospital, a residential area or a rural community. A cooperative IoT platform for rural healthcare monitoring has been found to be energy efficient [24].

v. Wearable device access: Various non-intrusive sensors have been developed for a diverse range of medical applications [25], particular for WSN-based healthcare services. Such sensors are prospective enough to deliver the same services through the IoT. On the other hand, wearable devices can come with a set of desirable features appropriate for the IoT architecture.

vi. Semantic medical access: The use of semantics and ontologies to share large amounts of medical information and knowledge has been widely considered [26].The wide potential of medical semantics and ontologies has received close attention from designers of IoT-based healthcare applications.

vii. Indirect emergency healthcare: There are some indirect emergency situations where healthcare issues are involved, including adverse weather conditions, transport (aviation, ship, train, and vehicle) accidents and earthen site collapses, among others. Therefore, a service called indirect emergency healthcare (IEH) can offer a bundle of solutions, such as information availability [2,27].

viii. Embedded gateway configuration: A configured gateway service, or embedded gateway configuration (EGC) service, connects network nodes to patients who are connected with the internet and all medical equipment. It requires some common integration features depending on the specific purpose of the deployed gateway [2].

ix. Embedded context prediction: One of the main issues with exbedded context prediction is the framework that all third-party developers may have to build with suitable mechanisms, that we called suitable mechanisms service [2].

Such a framework was developed by Mantas et al. [27] in the context of ubiquitous healthcare.

x. Early intervention/prevention: IoT may provide a way to monitor human activities and well-being, such as monitoring everyday activities and reporting them to the hospital or family members. In this way, IoT devices can be used for early intervention and prevention.

9.3.2 IoT Healthcare Applications

The following applications were selected by the incidence of population diseases, from the published paper of DGS (DirecçãoGeral de Saúde - Portugal) [28], and the need of an urgent responses from the medical community. Another source for analyzing the need of applications in medical environments is the HINtelligence report, 2015.

i. Diabetes prevention: "The term 'diabetes mellitus' describes a metabolic disorder of multiple etiology characterized by chronic hyperglycemia with disturbances of carbohydrate, fat and protein metabolism resulting from defects in insulin secretion, insulin action, or both. The effects of diabetes mellitus include long-term damage, dysfunction and failure of various organs" [1]. Blood glucose monitoring can prevent all the risks that this disorder may bring to patients by monitoring individual patterns of blood glucose and helping patients to plan their meals, activities and medication times.

ii. Electrocardiogram monitoring: According to one report [29], 30% of all deaths are related to circulatory system problems, such as arrhythmias, myocardial ischemia or prolonged QT intervals. Thus, it is important to monitor vital signals with an electrocardiogram (ECG)—the electrical activity of the heart recorded by electrocardiography includes the measurement of the heart rate and the determination of the rhythm as well as the diagnosis of arrhythmias, myocardial ischemia and prolonged QT intervals. Indeed, IoT-based applications for ECG monitoring have the potential to give maximum information and deliver information to medical staff [30].

iii. Blood pressure monitoring: This is a part of the prevention of circulatory system problems; therefore, IoT-based applications can remotely control communication between a health post and the health center [2].

iv. Body temperature monitoring: Homeostasis is how the human body manages a multitude of highly complex interactions to maintain balance or return systems to functioning within a normal range, like body temperature. Monitoring of this variable is an essential part of healthcare services because body temperature is a decisive vital sign. Using a body temperature sensor that is embedded in a TelosB device allows the sensor to retrieve body temperature variations and report them to a temperature measurement system based on a home gateway over the IoT [2].

v. Oxygen saturation monitoring: Blood oxygen saturation can be measured with a pulse oximetry, a non-invasive and non-stop monitoring system. The integration of a pulse oximetry with an IoT-based application can support oxygen saturation monitoring [6].

vi. Rehabilitation system: One of the main problems identified in the Portuguese population is aging and related medical issues, such as cerebrovascular accidents, which lead patients to enter rehabilitation clinics. In Fan et al. [31], an ontology-based automating design method for IoT-based smart rehabilitation systems is proposed to mitigate the problems previously described.

vii. Medication management: One of the main problems in public health and a huge financial burden is medication. IoT provides a new tool to resolve this issue [2].

viii. Wheelchair management: Smart wheelchairs with full automation for people with disabilities, use IoT to make adjustments such as acceleration in the workplace [2].

An framework that can be applied to a real application is the Remote Monitoring and Management Platform of Healthcare Information (RMMP–HI) [32]. This platform can provide monitoring and management of these lifestyle diseases to reach the purpose of prevention and early detection. Body medical sensors can register, delete and update data throughout an IoT-based network, then collect medical information from the body and send it to a data sharing center that propagates data to medical staff or hospital facilities based on rules, such as urgent notice derived to a hospital.

9.4 Detection of an Ambulance on Toll Roads: A Case Study of MIoT Application

9.4.1 Transportation Management System

A transportation management system, or TMS, is a platform that is designed to streamline the shipping process. It is a subset of supply chain management concerning transportation solutions. A TMS allows shippers to automate the processes they have in place and receive valuable insights to save time and reduce spending on future shipments.

Distribution companies, e-commerce organizations and anyone else who moves freight on a regular basis realizes there are many moving parts to the shipping process, both literally and figuratively. From quoting to delivery, those shipping freight are almost always looking for ways to optimize spending and improve processes. Thanks to TMS, shippers have a solution on their side to do just that [33].

9.4.1.1 Importance of Having a TMS

TMSs play a central role in supply chains, affecting every part of the process—from planning and procurement to logistics and life cycle management. The broad and deep visibility afforded by a powerful system leads to more efficient transportation planning and execution, which results in higher customer satisfaction. That, in turn, leads to more sales, helping businesses grow. With such a dynamic global trade environment that we live and transact in, it is important to have a system that will allow you to successfully navigate the complicated processes around trade policies and compliance [34].

9.4.1.2 Who Uses a TMS?

TMSs are primarily used by businesses that need to ship, move and receive goods on a regular basis, including the following:

- Manufacturers
- Distributors
- E-commerce companies
- Retail businesses
- Companies that provide logistics services, such as third-party and fourth-party logistics (3PL and 4PL) companies and logistics service providers (LSPs)

9.4.1.3 Plan, Execute and Optimize for Timely Delivery of Goods

A TMS can help any business plan, execute and optimize the physical movement of goods.

i. Planning: A TMS helps the business select the optimal mode of shipment and the best carrier, based on cost, efficiency and distance, including optimizing multi-leg carrier routes. A strong TMS can provide visibility into every stage of the supply chain, and together with global trade management functionality, it can provide information on trade and tariffs, and if there are any potential delays that may happen because of customs and other trade regulations.

ii. Execution: The execution features of TMSs vary widely but can include matching loads and communicating with carriers, documenting and tracking shipments, and assisting with freight billing and settlement. Some advanced TMS solutions also provide track-and-trace services—enabling real-time information exchange among carriers, distributors, warehouses and customers. Such advanced systems may also have the functionality to handle complex international logistics, including providing proper import and export documentation, making sure shipments are trade compliant.

iii. Optimization: TMS optimization capacities usually include the ability to measure and track performance with reports, dashboards, analytics and transportation intelligence [34].

9.4.1.4 The Future of TMSs

Customer expectations keep rising, not only for on-time deliveries but also for two-day and even same-day deliveries, with real-time updates provided throughout the shipment process. Ever-changing global trade regulations are also forcing supply chains to innovate to keep pace, often by investing in a TMS.

TMSs must become more robust and feature-rich, providing faster responses to consumers and more detailed information to businesses. Machine learning enables TMSs to be more intelligent, providing better recommendations and more accurate predictions.

Companies can choose to integrate their transportation and global trade management systems with emerging technologies to further improve visibility and offer better customer service. Some innovative technologies that are currently available include:

    i. IoT fleet monitoring: IoT devices and sensors make real-time fleet monitoring commonplace, including in-transit visibility of driving conditions, routes and assets. Companies can lower their fuel and maintenance costs, as well as reduce delays and improve driver safety.

    ii. Digital assistants: Digital assistants are often called chatbots, and they offer immediate conversational responses to shipment information, leading to higher customer satisfaction.

    iii. Adaptive intelligence and machine learning: By applying machine learning to historical data and trends, TMSs are able to predict transit time more accurately, plan capacity, identify at-risk shipments (e.g., goods that are about to expire, time- or temperature-sensitive products), and much more. Enhanced AI will also enable your TMS to provide more accurate and informed recommendations, such as alternate delivery routes during high traffic periods.

    iv. Blockchain: Blockchains are now being utilized to build complex integrations among shippers, customers and carriers. Applications such as intelligent track and trace increase transparency and trace ability across your supply chain but still ensure accurate and secure information.

    v. Cold chain management: Another blockchain solution available in TMSs is cold chain management, which is useful when different temperatures need to be maintained at various checkpoints along the supply chain. For instance, perishable or temperature sensitive materials and products might need to be kept at a cool temperature in the truck but a slightly higher temperature on store shelves. With cold chain management, the temperature can be monitored across the supply chain, with real-time information provided to the business and the regulators at the country of origin [34].

9.4.2 Radio-Frequency Identification (RFID)

RFID uses electromagnetic fields to automatically identify and track tags attached to objects. The tags contain electronically stored information. Passive tags collect energy from a nearby RFID reader's interrogating radio waves. Active tags have a local power source (e.g., battery) and may operate hundreds of meters from the RFID reader. Unlike a barcode, the tags do not need to be within the line of sight of the reader, so they may be embedded in a tracked object. RFID is one method of automatic identification and data capture (AIDC) [35].

    RFID tags are used in many industries. For example, an RFID tag attached to an automobile during production can be used to track its progress through the assembly line; RFID-tagged pharmaceuticals can be tracked through warehouses; and implanting RFID microchips in livestock and pets enables positive identification of animals.

9.4.2.1 RFID Design

9.4.2.1.1 Tags

A RFID system uses tags, or labels attached to the objects to be identified. Two-way radio transmitter-receivers, called interrogators or readers, send a signal to the tag and read its response [36].

RFID tags can be passive, active or battery-assisted passive (BAP). An active tag has an on-board battery and periodically transmits its ID signal. A BAP has a small battery on board and is activated when in the presence of an RFID reader. A passive tag is cheaper and smaller because it has no battery; instead, the tag uses the radio energy transmitted by the reader. However, to operate a passive tag, it must be illuminated with a power level roughly 1,000 times stronger than its signal transmission. That makes a difference in interference and in exposure to radiation. Tags may either be read only, having a factory-assigned serial number that is used as a key into a database, or may be read/write, where object-specific data can be written into the tag by the system user. Field programmable tags may be write-once, read-multiple, or "blank" tags, which may be written with an electronic product code by the user.

RFID tags contain at least three parts: an integrated circuit that stores and processes information and that modulates and demodulates RF signals; a means of collecting DC power from the incident reader signal; and an antenna for receiving and transmitting the signal. The tag information is stored in a non-volatile memory. The RFID tag includes either fixed or programmable logic for processing the transmission and sensor data, respectively.

9.4.2.1.2 Readers

An RFID reader transmits an encoded radio signal to interrogate the tag. The RFID tag receives the message and then responds with its identification and other information. This may be only a unique tag serial number, or it may be product-related information such as a stock number, lot or batch number, production date or other specific information. Since tags have individual serial numbers, the RFID system design can discriminate among several tags that might be within the range of the RFID reader and read them simultaneously.

RFID systems can be classified by the type of tag and reader. A passive reader active tag (PRAT) system has a passive reader that only receives radio signals from active tags (battery operated, transmit only). The reception range of a PRAT system reader can be adjusted from 1–2,000 feet (0–600 m) [37], allowing flexibility in applications such as asset protection and supervision.

An active reader passive tag (ARPT) system has an active reader, which transmits interrogator signals and also receives authentication replies from passive tags. An active reader active tag (ARAT) system uses active tags awoken with an interrogator signal from the active reader. A variation of this system could also use a BAP tag, which acts like a passive tag but has a small battery to power the tag's return reporting signal.

Fixed readers are set up to create a specific interrogation zone that can be tightly controlled. This allows a highly defined reading area for when tags go in and out of the interrogation zone. Mobile readers may be handheld or mounted on carts or vehicles.

9.4.2.1.3 Uses

The RFID tag can be affixed to an object and used to track and manage inventory, assets, people, etc. For example, it can be affixed to cars, computer equipment, books or mobile phones.

RFID offers advantages over manual systems or use of bar codes. The tag can be read if passed near a reader, even if it is covered by the object or not visible. The tag can be read inside a case, carton, box or other container, and unlike barcodes, RFID tags can be read hundreds at a time. Bar codes can only be read one at a time using current devices

Transportation and Logistics

Yard management, shipping and freight, and distribution centers use RFID tracking. In the railroad industry, RFID tags mounted on locomotives and rolling stock identify the owner, identification number, and type of equipment and its characteristics. This can be used with a database to identify the lading, origin, destination, etc., of the commodities being carried [38].

In commercial aviation, RFID is used to support maintenance on commercial aircraft. RFID tags are used to identify baggage and cargo at several airports and airlines [35,39]. Some countries use RFID for vehicle registration and enforcement [40]. RFID can help detect and retrieve stolen cars [41,42]. RFID E-ZPass readers are attached to the pole and mast arm (towards right side) used in traffic monitoring in New York City. RFID is used in intelligent transportation systems. In New York City, RFID readers are deployed at intersections to track E-ZPass tags as a means for monitoring the traffic flow. The data are fed through the broadband wireless infrastructure to the traffic management center to be used in adaptive traffic control of the traffic lights [43].

In hose stations, the RFID antenna in a permanently installed coupling half (fixed part) unmistakably identifies the RFID transponder placed in the other coupling half (free part) after completed coupling. When connected, the transponder of the free part transmits all important information contactlessly to the fixed part. The coupling's location can be clearly identified by the RFID transponder coding. The control is enabled to automatically convey fluids.

In the automotive industry RFID is used to track and trace test vehicles and prototype parts (Project Transparent Prototype).

9.4.3 The Proposed Methodology

Continuous population growth became a great challenge for all TMSs. With the rapid growth of the Indian economy, the development of TMS makes use of an intelligent transportation system, which helps to detect an ambulance on a toll road and thus helps to release it as soon as possible from the traffic by RFID process for detecting the ambulance, which makes electrostatic coupling usage in the portion of radio frequency in electromagnetic spectrum for identifying objects specially.

The radio waves that have been produced are used for transmitting data from the source to the receiver, which then transmits the information to the RFID computer program. In the field of surveillance for traffic management, it is an important tool and has been considered one of the fastest growing technologies. The aim is to detect the ambulance with accuracy. The input can be obtained either in the form of a still image or video footage from the surveillance camera, where it can be processed either at the color level or at the grayscale level.

Compared to the color level, grayscale processing is considered to be a more efficient way to obtain a "+" sign accurately because the color level does not support processing of large data. Less data is handled at the grayscale level; thus, it facilitates faster processing compared to the color level; 98.5% accuracy has been proposed for the method depending on information on texture [44].

In the case of hazy images with a "+" sign, the objective of the algorithm is to obtain a clearer image, which is difficult to solve. Thus, the technique of horizontal and vertical projection of images on the X and Y axis is implemented to obtain a clear image. From video footage, the image is captured and is then passed through a Gaussian filter and histogram equalization, and morphological operations are

FIGURE 9.5
Original image (left) vs Gaussian filtered image (right).

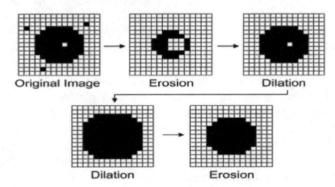

FIGURE 9.6
The technique of dilation and erosion for extraction of symbols.

performed for extracting the "+" sign. A distinction has been drawn between the original and the Gaussian filtered image in Figure 9.5.

Dilation and erosion, which are types of morphological operations, are specially used to obtain the exact symbol of the ambulance, where the sign of the ambulance is extracted as a complex image by cropping the image from the rest of the vehicle images. This technique is applied when vehicles are not clearly visible. These techniques are further illustrated in Figure 9.6.

Not only tilted images but also distorted images can be resolved by using image till correction and gray-level technique. In most cases, an ambulance is identified by its "+" sign, which is easily recognized by humans but not by machines as the machines understand only binary numbers, 0 and 1.

The image is only a grey picture, which has been defined as a two-dimensional function $f(x, y)$, where x and y are spatial coordinates and f is an intensity of that point. Hardware parts consist of a camera image processor, a camera trigger, communications and a storage unit. Software parts consist of an input image or video footage that the user has sent.

It is also necessary to ensure the invariance of the system toward the condition of light; for example, a normal camera is not suitable for capturing pictures in dark conditions as it follows the spectrum of visible light.

Color is defined by the Law of Nature with the perception of light, which delivers the cone cell stimulus in the human eye through electromagnetic radiation by light spectrum. Wavelength and intensity are the characterization of electromagnetic radiation. When the wavelength is between 390 nm and 700 nm, the visible spectrum of light is obtained. The ability to distinguish colors depends on the varying sensitivity of light of various wavelengths. The retina of the human eyes is sensitive to three color receptor cells, namely short-wavelength cones or S-cones or blue cones, which are sensitive to blue color of wavelength 450 nm; medium-wavelength cones or M-cones or green cones, which are sensitive to green colors of wavelength 540 nm; and long-wavelength cones or L-cones or red cones, which are sensitive to greenish-yellow color of wavelength 570 nm.

The mixing of two or more colors creates an additive color. Red, green and blue are considered to be additive colors and are used in projectors and computer terminals. The colors, which absorb some part of the wavelength of light, make subtractive colors, which can be obtained by adding pigment or ink, and thus another color of light reaches the human eye. This additive and subtractive color combination is illustrated in Figure 9.7.

An RGB color space is defined as the space of additive colors, which is based on the RGB color model that produces a convenient color and consists of three chromaticity-like red, green and blue colors and thus produces another chromaticity, which is supplemented on a triangle depending on the three additive colors. A white point chromaticity and gamma correction curve is required for the overall specification of RGB color space [45]. Another subtractive color model is the CMYK model, which consists of four colors, namely cyan, magenta, and yellow, and "K" is for key that indicates black. The comparisons between RGB and CMYK are not possible as the reproduction of color technologies as well as the properties are totally different. Further, this comparison is diagramatically explained in Figure 9.8.

A digital image is defined as a numerical representation of 2D images, which is basically referred to as a raster or bitmapped image that has a finite set of binary values known as pixels or picture elements, which consists of a defined set of rows or columns.

Pixels can also be defined as the smallest part of an individual element of an image that holds a specific value that defines the brightness at specific point of any given color. The location of pixel in the computer's memory is in the form of a raster image or as the raster map, which is in the form of a 2D array of integers.

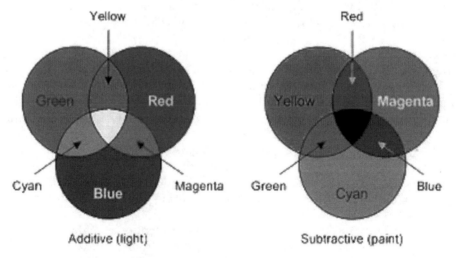

FIGURE 9.7
Additive and subtractive color combinations.

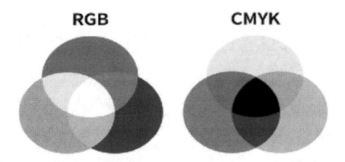

FIGURE 9.8
Comparison between RGB and CMYK color.

The creation of raster images is done through different input devices and also through various techniques, namely digital cameras and scanners that can be codified through erratically non-image data like mathematical functions or 3D models, which is a widened part of computer graphics. Mathematical geometry is the result of vector image.

According to mathematics, vector is defined as a point that consists of both length and direction. Standard Eastern Automatic Computer (SEAC) at NIST has scanned, stored as well as recreated the first picture on pixel.

For digital image, pixel can be defined as a physical point or a small addressable element of a raster image that can be arranged on a 2D regular grid. By implementing this arrangement, commonly used operations can be performed by considering individual pixels at a time [46].

Now, a digital image can be represented based on three scales. These are as follows:

1. Binary scale image
2. Gray scale image
3. Color scale image

Color scale image has two types:

- RGB mode
- CMYK mode

A binary image can be defined as a digital image that can have two values for individual pixels. To represent a binary image, only two colors can be used: black and white [47].

For objects, the color that is used is known as the foreground color whereas the leftover part of image is known as the background color. Another name for binary images are bi-level or two-level images, which signifies that each pixel can be used to store a single bit (i.e., 0 and 1). An arrangement of bitmaps or bits packed like an array can be used to store a binary image.

A gray scale image can be defined as an image where the value of each pixel can be stored as a single sample that only carries the intensity information. The word "gray" is defined from the combination of black and white, which ultimately yields a gray color, and "grayscale" means the combination of shades of gray, which can vary from the range of weakest intensity to strongest intensity (i.e., from black to white) [48].

The measurement for the intensity of light for each pixel on a single band of the electromagnetic spectrum results in a grayscale image, and only a particular frequency gets captured in a monochromatic way. The intensity is always expressed between the range of minimum and maximum inclusively, which can be represented from black (0, total absence) to white (1, total presence), where fractional values are accepted.

Thus, a digital color image can be defined as an image where information on colors for each pixel gets included. There are three values (or channels) for each pixel in a color image that compute the intensity as well as the chrominance of light. The information on brightness for each band of spectra gets stored in the digital image. Dividing R, G, B values by 255 for changing the range to 0.1–0.255, where R' = R/255, B' = B/255, G' = G/255. Black key (K) color can be calculated from red (R'), green (G') and blue (B') colors, which is K = 1 max (R',G',B').

Some pixels consist of two vertical as well as two horizontal neighbors. The pixels having these sets are known as 4-neighbors, which are at a Euclidean distance from each pixel. If some specifically mentioned criteria of similarity is satisfied by both the neighbors and gray scale level, then we say that there exists a connection between two pixels. Every operation of image can be defined by a convolution matrix that defines the over-ridation of one pixel due to neighboring pixels by the convolution process. An image filter is defined as a procedure that can change the emergence of a particular image or a part of that image by either changing the shades or the colors of pixels [49].

The use of a filter is either to increase the brightness or contrast of the image or to provide a broad variety of textures or tones and add special effects to a particular picture. There are two types of filters, namely mean filter and Gaussian blur filter. Mean filter, also known as smoothing, can be defined as a straightforward, instinctive and simple method for decreasing the value of intensity variation that links one pixel to other. It lessens noise in images. The main idea for mean filtering is replacing the value of each image by the mean value of the neighboring pixels including that pixel. The distinction between mean and Gaussian filtering is depicted in Figure 9.9.

A Gaussian blur, or Gaussian smoothing, un-focuses part of image with the help of the Gaussian function. It is used widely on graphics software for reducing the noise in an image and also to reduce detail. The process of convolution of image along with the Gaussian function is called 2D Weierstrass transform.

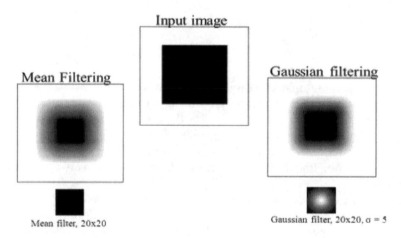

FIGURE 9.9
Mean vs Gaussian filtering.

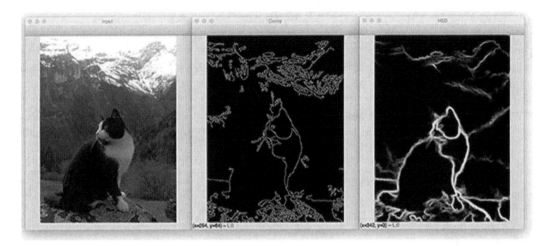

FIGURE 9.10
Edge detection of an Image.

As Gaussian blur is used for reducing the components of high frequency in an image, it is considered to be a low pass filter. For detecting the edges of an image, Sobel operator is used. It is also known as the Sobel-Feldman operator or Sobel filter, which is used for processing the image and vision in the computer for detecting the edges on an image. The edge detection of an image is illustrated in Figure 9.10.

Because the Sobel-Feldman operator uses convolving a compact, divisible as well as integer-valued filter on horizontal along with vertical direction, it is inexpensive regarding computations. Comparatively, gradient approximation produces a crude outcome, particularly in variations with high frequency. Thus, the operator is composed of 3 × 3 convolution kernel pairs [50].

Morphology is defined as a broadened set for processing operations on an image depending on the shapes and produces the output image the same as the input image. Erosion is a fundamental operation, explained only for binary images that contract objects in binary images. Dilation is another morphological operation that expands the objects for binary images [51].

Region growing can be defined as a straightforward region-based segmentation of images that involves selecting initial seed points and checks whether the neighbors of pixels could be added or not and thus is iterated the same as a general algorithm for data clustering [52]. The advantages of region growing are as follows:

1. It separates regions that have the same properties.
2. It provides the original image with better segmentation and with clear edges.
3. Determination of seed point along with criteria is possible.
4. Choosing multiple criteria at once is possible.

The disadvantages are as follows:

1. It is computationally expensive.
2. It is noise sensitive.
3. It is a local method, and the global view is absent.

The vertical projection can be defined as a graph that is used to represent the magnitude of the overall image with respect to the Y-axis, and horizontal projection is used to represent the mapping in the X-axis. The Hough-transform is simply a technique for the extraction of features that can be used in the analysis of image, computer vision and digital image processing. Its purpose is finding the imperfections in objects of a certain class with the help of a voting procedure.

The process of segmenting the characters into individual letters is called fuzzy spike. The continuous region in 1D on non-zero histogram is called spike [53]. Fuzzy spike can also be defined as a cluster of histogram values having the values at intermittent regions as zero.

Bilinear interpolation slogs by interpolating different color values in pixels by introducing continuous transition to form an output where the original image is arranged in a discrete manner. Bicubic interpolation often yields a better output with a very small increase in computational complexity [54]. These interpolations are visible in Figure 9.11.

The above procedures are the existing methodologies as well as algorithms for vehicles used to recognize the ambulance sign. The study of the different procedures for determining the sign on the ambulance provides the merits, drawbacks, computational power and accuracy. It encourages us to develop advanced ways to determine the sign on the ambulance [55].

The RFID process can help solve traffic problems such as determining the priority of different kinds of vehicles and identifying the traffic density on a road; an RF reader can be installed at the intersection of two roads [56]. RFID is a technology where digital data are encoded inside RFID tags as well as on smart labels that will be captured by readers through radio waves. The tags that are captured by the device will store its data in a database. One of the advantages is that the data that are stored inside RFID tags can be read outside of view. RFID resides under a group of technologies called AIDC, which identifies objects, collects data and enters the data into a computer with no or very little human involvement. It uses radio waves to accomplish this.

RFID is a popular wireless inauguration systems. RFID tags come in various shapes and sizes and are read by an RF reader. If an ambulance is on a toll road, RFID can be used to control traffic signals. The signal is set to green for a predefined time. After the ambulance passes, the original signal flow can return.

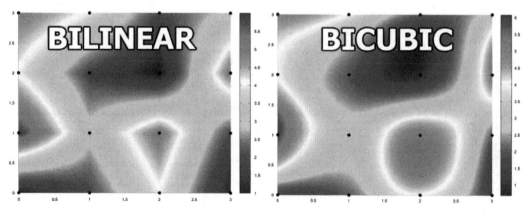

FIGURE 9.11
Bilinear vs bicubic interpolation.

There are two categories of RFID tags: passive tags and active tags. Tags are composed of three parts: antenna, some parts of encapsulation and a semiconductor chip. For storing data and performing specific tasks, RFID chips are used. Depending on the designs, the chips can be read only (RO), write-once read many (WORM) or read-write (RW).

The antenna is used for absorbing radio waves from the signal that has been received from the reader as well as for sending and receiving data. The performance of RFID passive tags largely depends on the size of antenna; the larger the antenna, the more energy is collected and returned. The shape of the antenna is also essential for the performance of the tags. Low-frequency (LF) and high-frequency (HF) antennas are coiled because the frequencies are magnetic in nature [57].

Both the antenna as well as the chip are attached to a substrate that holds the tag's pieces together. These tags also include protective materials that hold the pieces together, guarding them from different environmental conditions. The scope of passive tags is wide. The reader sends electromagnetic waves to produce the tag antenna's current; thus, the antenna throws back the information that has been stored inside it. It needs to be powered up before transmitting data.

The active tag consists of both a microchip and an antenna. The size of the chip is large, and the capabilities are greater than passive tags. The active tag includes a battery that is used as a power source (internal) for operating the microchip's circuitry as well as broadcasting received information to the reader. The cost and range of these kinds of tags are comparatively more than passive tags. There are two supplementary components of active tags that make them differ from passive tags: onboard power supply and onboard electronics.

The onboard power supply is mainly a battery, but it can be solar. The built-in power helps to transmit data from the tag to the reader without drawing any power from the reader. The onboard electronics comprises sensors, microprocessors and input or output ports. The tag's onboard power source provides power to all of the components [57].

There are four memory banks in each chip: EPC, TID, user and reader. Each memory bank has information about the tagged item.

EPC (electronic product ode): This code is the syntax for distinctive identifiers allocated to physical objects, unit loads, locations or recognizable entities that play some role in the operations of the business. There are multiple representations of EPC that include binary forms and can be used in RFID tags. They also include text forms that can be used for sharing data among systems. These radio waves are used for capturing distinct identifiers at very high rates. EPC can be changed easily. It is a number that has been written on a passive RFID transponder by the product's manufacturer. It can be used for identifying a product, the product's category and the manufacturer of the product.

TID (transponder ID): This ID is used for identifying the type of chip, the custom commands as well as some optional features that the chip supports. Without distinct identification of chips, this can be possible; thus, serialization of the TID number is not required. When a distinct serial number is added, it identifies a unique chip. The TID number starts with an eight-digit allocation-class (AC) identifier. It explains the mechanism that guarantees the uniqueness of TID numbers. It cannot be changed easily. It is a distinct number that has been written in the transponder's microchip by the microchip's manufacturer. It is not readable by RFID readers, but there is a possibility of creating a system that authenticates a tag after cross-checking TID and EPC.

Based on ranges of frequency, there are three kinds of RFID tags: low-frequency, high-frequency and ultra high frequency. The IR transmitter sender and IR receiver communicate with each other by means of the line-of-vision propagation method. There is an eight-digit distinct serial number on the RFID tag on the ambulance. The RFID receiver reads the tag serial number by means of electromagnetic waves when the ambulance travels through the toll road. RFID tags are embedded within the databases and are then matched with the tag from the ambulance.

When the ambulance passes through the toll road, the tag on the ambulance is displayed on the screen. This will reduce the need for ambulances to stand by on toll roads. The accuracy of the RFID process is more than a camera's ability to determine ambulances on toll roads. Thus, we can use this process in more advanced ways for future enhancements. It can send messages to doctors so that they can prepare for patients before they arrive at the hospital. Thus, it can also provide for better and faster treatment of patients.

9.5 Conclusion

This chapter reviewed various aspects of MIoT. It began with a discussion on IoT, deeply studied its architecture and the protocol stack involved in it, and covered how IoT has been utilized in medical applications to evolve a new term "MIoT". The chapter further studied the various applications and services provided by MIot. The needs and future aspects of MIot also were discussed. Finally, the chapter presented a case study to study how RFID can be utilized to identify ambulances at toll booths.

References

1. T. N. Gia, A.-M. Rahmani, T. Westerlund, P. Liljeberg, H. Tenhunen, Fault tolerant and scalable IoT-based architecture for health monitoring. In: *IEEE Access* (2015).
2. S. M. Riazul Islam, D. Kwak, Md H. Kabir, M. Hossain, K.-S. Kwak, The Internet of Things for health care: a comprehensive survey. In: *IEEE Access* (2015).
3. G. Santucci and S. Lange, "Internet of things in 2020: A Roadmap for the Future", European Commission (DG INFSO) and European Technology Platform on Smart Systems Integration (EPoSS) Workshop Report, (2008).
4. O. Vermesan, P. Friess, P. Guillemin, S. Gusmeroli, H. Sundmaeker, A. Bassi, I.S. Jubert, M. Mazura, M. Harrison, M. Eisenhauer and P. Doody, "Internet of Things Strategic Research Roadmap", Internet of Things - Global Technological and Societal Trends, River Publishers, ISBN - 9788792329677, pp. 9–52.
5. L. Atzori, A. Iera, G. Morabito, "The Internet of Things: A survey," *Computer Networks*. vol. 54, pp. 2787– 280 (2010).
6. H. A. Khattak, M. Ruta, E. Di Sciascio, CoAP-based healthcare sensor networks: A survey. In: *Proc. 11th Int. Bhurban Conf. Appl. Sci. Technol. (IBCAST)*, pp. 499–503 (2014).
7. X. M. Zhang, N. Zhang, "An open, secure and exible platform based on Internet of Things and cloud computing for ambient aiding living and telemedicine,"' *Proc. Int. Conf. Comput. Manage. (CAMAN)*, 2011, pp. 1–4.
8. S. Imadali, A. Karanasiou, A. Petrescu, I. Sifniadis, V. Veque, P. Angelidis, "eHealth service support in IPv6 vehicular networks,"' In *Proc. IEEE Int. Conf. Wireless Mobile Comput., Netw. Commun. (WiMob)*, 2012, pp. 579–585.

9. A. J. Jara, M. A. Zamora, A. F. Skarmeta, "Knowledge acquisition and management architecture for mobile and personal health environments based on the Internet of Things,"' In *Proc. IEEE Int. Conf. Trust, Security Privacy Comput. Commun. (TrustCom)*, 2012, pp. 1811–1818.

10. C. Doukas, I. Maglogiannis, "Bringing IoT and cloud computing towards pervasive healthcare,"' In *Proc. Int. Conf. Innov. Mobile Internet Services Ubiquitous Comput. (IMIS)*, 2012, pp. 922–926.

11. M. S. Shahamabadi, B. B. M. Ali, P. Varahram, A. J. Jara, "A network mobility solution based on 6LoWPAN hospital wireless sensor network (NEMO-HWSN)."' In: *Proc. 7th Int. Conf. Innov. Mobile Internet Services Ubiquitous Comput. (IMIS)*, 2013, pp. 433–438.

12. M. F. A. Rasid et al., "Embedded gateway services for Internet of Things applications in ubiquitous healthcare,"' In *Proc. 2nd Int. Conf. Inf. Commun. Technol. (ICoICT)*, 2014, pp. 145–148.

13. Z. Shelby, C. Bormann, *6LoWPAN: The Wireless Embedded Internet*, 1st ed. London, U.K.: Wiley, 2009.

14. N. Bui, N. Bressan, M. Zorzi, "Interconnection of body area networks to a communications infrastructure: An architectural study," In *Proc. 18th Eur. Wireless Conf. Eur. Wireless*, 2012, pp. 1–8.

15. L. You, C. Liu, S. Tong, "Community medical network (CMN): Architecture and implementation,"' In *Proc. Global Mobile Congr. (GMC)*, 2011, pp. 1–6.

16. P. Swiatek, A. Rucinski, "IoT as a service system for eHealth,"' In *Proc. IEEE Int. Conf. eHealthNetw., Appl. Services (Healthcom)*, 2013, pp. 81–84.

17. M. Diaz, G. Juan, O. Lucas, A. Ryuga, "Big data on the Internet of Things: An example for the e-health," In *Proc. Int. Conf. Innov. Mobile Internet Services Ubiquitous Comput. (IMIS)*, 2012, pp. 898–900.

18. X. Wang, J. T. Wang, X. Zhang, J. Song, "A multiple communication standards compatible IoT system for medical usage."' In *Proc. IEEE Faible Tension FaibleConsommation (FTFC)*, 2013, pp. 1–4.

19. B. Xu, L. D. Xu, H. Cai, C. Xie, J. Hu, F. Bu, "Ubiquitous data accessing method in IoT-based information system for emergency medical services," *IEEE Trans. Ind. Informat.*, vol. 10, no. 2, pp. 1578–1586, May 2014.

20. R. S. Istepanian, N. Y. Philip, A. Sungoor, The potential of Internet of m-health Things "m-IoT" for non-invasive glucose level sensing. In: *Conf. Proc. IEEE Eng. Med. Biol. Soc.* (2011).

21. R. S. H. Istepanian, E. Jovanov, Y. T. Zhang, Guest editorial introduction to the special section on m-health: Beyond seamless mobility and global wireless health-care connectivity. In: *IEEE Trans. Inf. Technol. Biomed.* pp. 405–414 (2004).

22. Group, I. E. W., Guidance for industry-E6 good clinical practice: Consolidated guidance. In: *U.S. Dept. Health Human Services, Food Drug Admin* (1996).

23. A. J. Jara, F. J. Belchi, A. F. Alcolea, J. Santa, M. A. Zamora-Izquierdo, A. F. Gomez-Skarmeta, A pharmaceutical intelligent information system to detect allergies and adverse drugs reactions based on Internet of Things. In: *Proc. IEEE Int. Conf. Pervasive Comput. Commun. Workshops (PERCOM Workshops)*. pp. 809–812 (2010).

24. V. M. Rohokale, N. Prasad, R. Prasad, A cooperative Internet of Things (IoT) for rural healthcare monitoring and control. In: *Proc. Int. Conf.WirelessCommun., Veh. Technol., Inf. Theory Aerosp.Electron. Syst. Technol. (Wireless VITAE)*, pp.1–6 (2011).

25. W.-Y. Chung, Y.-D. Lee, S.-J. Jung, A cooperative Internet of Things (IoT) for rural healthcare monitoring and control. In: *A wireless sensor network compatible wearable u-healthcare monitoring system using integrated ECG, accelerometer and SpO2*. pp. 1529–1532 (2008).

26. A. Burgun, G. Botti, M. Fieschi., P. Le Beux, Sharing knowledge in medicine: Semantic and ontologic facets of medical concepts. In: *Proc. IEEE Int. Conf. Syst., Man, Cybern. (SMC)*. pp. 300–305 (1999).

27. G. Mantas, D. Lymberopoulos, N. Komninos, A new framework for ubiquitous context-aware healthcare applications. In: *Proc. 10th IEEE Int. Conf. Inf. Technol. Appl. Biomed. (ITAB)*. pp. 1–4 (2010).

28. F. George, Causas de Morteem Portugal e DesafiosnaPrevenção. *DGS* (2012).

29. D. G. da Saúde, A Saúde dos Portugueses. Perspetiva 2015 (2015).

30. P. K. Dash, Electrocardiogram monitoring. *Indian J. Anaesthesia*, vol. 46 pp. 251–260 (2002).

31. Y. J. Fan, Y. H. Yin, L. Xu, Y. Zeng, F. Wu, IoT-based smart rehabilitation system. In: *IEEE Trans. Ind. Informat*. pp. 1568–1577 (2014).

32. W. Zhao, C. Wang, Y. Nakahira, Medical Application OnIoT. In: *International Conference on Computer Theory and Applications (ICCTA)*. pp. 660–665 (2011).

33. C. H. Robinson, "What is a transportation management system (TMS)?", *Freightquote*, 2020, https://www.freightquote.com/define/what-is-transportation-management-system-tms/ [online].

34. Oracle.com, "What is a transportation management system?", *Oracle*, 2020, https://www.oracle.com/applications/supply-chain-management/what-is-transportation-management-system.html [online].

35. MHI, "Automatic identification and data collection (AIDC)", *Wayback Machine*, Archived May 5, 2016.

36. Usmijr, "RFID-Tag". *Behance*. Retrieved 2018-07-15.

37. G. Phillips, "How Does RFID Technology Work?". *Technology Explained*, (2017). https://www.makeuseof.com/tag/technology-explained-how-do-rfid-tags-work/ [online]. Retrieved 2019-04-22.

38. Wayback machine, "AEI technology". *Softrail. Archived from the original on 2008-04-06*. Retrieved 2008-10-12.

39. Jetstar, "Qantas next generation check-in". *Qantas Airways Limited*. Retrieved 2010-12-27.

40. R. Wessel, "Bermuda's RFID vehicle registration system could save $2 Million/Year". Rfidjournal.com. 2007-05-18. Retrieved 2013-09-03.

41. Infineon Technologies, "Smart license may cut car theft". Rfidjournal.com. 2002-10-11. Retrieved 2013-09-03.

42. P. Samuel, "Mexico's electronic vehicle registration system opens with Sirit open road toll technology, Dec 29, 2009". *Tollroadsnews.com. Archived from the original on* 2013-07-03. Retrieved 2013-09-03.

43. ITS International, "New York's award-winning traffic control system". *ITS Int*. January–February 2013. Retrieved 2014-05-03.

44. J. Wolfgang, V. Goethe, "Gorthe's theory of colors", *Art, Philosophy*, 2013. [online] http://www.openculture.com/2013/09/goethes-theory-of-colors-and-kandinsky.html

45. D. Pascale, "A review of RGB Color Spaces … from xyY to R'G'B'", *The BabelColor Company*, Retrieved 2008-01-21.

46. R. Lyon. (2006). A brief history of 'pixel. *Proceedings of SPIE - The International Society for Optical Engineering.6069*. 10.1117/12.644941.

47. J. D. Foley, A. Van Dam (1982). *Fundamentals of Interactive Computer Graphics*. Reading, MA: Addison-Wesley. ISBN 0201144689.

48. J. D. Cook, "Three algorithms for converting color to grayscale", Posted on 24 August 2009. [online] https://www.johndcook.com/blog/2009/08/24/algorithms-convert-color-grayscale

49. O. Lecarme, K. Delvare (January 2013). The book of GIMP: A complete guide to nearly everything. *No Starch Press*. p. 429. ISBN 978-1593273835.

50. R. Brunelli, *Template Matching Techniques in Computer Vision: Theory and Practice*, Wiley, ISBN 978-0-470-51706-2, 2009.

51. R. Gonzalez (2018). *Digital Image Processing*. New York, NY: Pearson. ISBN 978-0-13-335672-4. OCLC 966609831.

52. S. Tania, R. Rowaida (2016). A comparative study of various image filtering techniques for removing various noisy pixels in aerial image. *Int. J. Signal Processing, Image Processing Pattern Recogn*. Vol. 9. 113–124. 10.14257/ijsip.2016.9.3.10.

53. A. Campbell The designer's Lexicon. *2000 Chronicle, San Francisco*. p 192.

54. H. Robert (2004). Exploring colour photography: A complete guide. *Laurence King Publishing.* ISBN 1-85669-420-8.

55. M. Ondrej (2007). "Algorithm & mathematical principles of automatic number plate recognition systems", B.SC.Thesis, *Faculty of Information Technology Department of Intelligent Systems,* Brno University of Technology.

56. The MV Act (Rule 50, 51 of MV Act), 1989 [online] http://www.htp.gov.in/Annexure-I.pdf

57. InvestmentKit, "Rules, format and example of number plates of vehicles in India", November 20, 2013. [online] https://www.investmentkit.com/articles/2013/11/rules-format-and-example-of-number-plates-of-vehicles-in-india

10

Diagnosis and Treatment of Cardiac Patients During the Pandemic Situation Due to COVID-19: Exploring Possible Application of MIoT

Sudip Chatterjee[1], Gour Sundar Mitra Thakur[2] and Shyamapriya Chowdhury[3]

[1]*Department of CSE, Amity University Kolkata, Kolkata, India*
[2]*Department of CSE, Dr. B.C. Roy Engineering College, Durgapur, India*
[3]*Department of Information Technology, Narula Institute of Technology, Kolkata, India*

CONTENTS

10.1 Introduction

The World Health Organization (WHO) published a report called "Global Health and Aging Report" [1] in 2010 in which they presented that the number of people older than 65 years was almost 524 million, and they also estimated that the number might increase up to 1.5 billion by 2050. Elderly people need some extra care due to their various difficulties such as inability to move, vulnerability to diseases (e.g., COVID-19) while they are exposed, and most importantly, illness due to suffering from different chronic diseases. In both developed and developing countries, cardiovascular disease (CVD) is the leading causes of mortality and morbidity [2]. According to the Global Burden of Disease Report, released in September 2017, CVD caused 1.7 million deaths, which is 17.8% in 2016, and it is the leading cause of death in India [3]. It takes 50% of heart attack cases

DOI: 10.1201/9780429318078-10

more than 400 minutes to reach the hospital versus the ideal window of 180 minutes, beyond which the damage is permanent [4]. So, it can be inferred from the findings that for CVD patients, especially elderly patients, it is very important to monitor their health and treat problems in time.

During this COVID-19 pandemic, monitoring or treating of CVD patients has become extremely difficult due to the threat of exposing patients, especially older patients, to health workers outside the home. Apart from this, non-availability/paucity of proper vehicles (e.g., ambulance) in this situation has also become a hurdle for providing prompt treatment to patients. In compliance with recent strategies planned by state governments and central governments, various government and private hospitals are accommodating COVID-19 patients, and as a consequence, the scope of treatment of CVD patients and other patients has been squeezed. Therefore, CVD patients are finding it very hard to get proper treatment or receive treatment in a timely manor. Older patients are facing more difficulties in this respect.

In this crucial condition, MIoT can play a pivotal role in providing the required treatment to CVD patients. Bhagchandani and Augustine [4] first captured medical data in real time using MIoT (smart sensors) and then computed the data in emergency situations. After collecting heart-related data from the input parameters, their proposed a system that triggers an alarm to notify a patient's relatives, doctors, ambulance, etc. The smart-based application displays heart rate and other related data of a patient; if the data deviate from pre-stored standard data, the software generates an alarm which reaches to the concerned person's smartphone with all the necessary data along with a graph. MIoT can be described as the application of well-recognized basics, principles, methods, techniques and concepts of internet-oriented approaches specifically for the medical and healthcare sectors [5]. An MIoT-based system means an efficient-enough serving network comprising healthcare resources and medical services connected through internet-based devices. MIoT uses apps and ICT (information and communication technology) to connect wearable sensors, patients, healthcare professionals and caregivers [6–8].

Singh et al. [6] presented a novel idea of amalgamating medical diagnostic instruments and their applications in the form of MIoT to connect to healthcare facilities to provide critical treatment for orthopedic patients. The authors explored the infrastructure, tools, devices and method of operation required for providing such diagnosis and treatment. Physical devices, such as channel devices, sensors and monitors, are used for data collection and for analyzing objects. The authors illustrate the process of data transformation to an understandable form, data processing and use of the cloud environment for computation on processed data. Lastly, they discuss advanced analysis of data for ultimate action to be taken as per the requirements of the patient. Joyia et al. [7] illustrated the research on MIoT in the healthcare sector published between 2012 and 2016. They also presented various challenges in the medical sector which should be overcome to build an efficient, secured system. The challenges include hardware and device implementation, software requirements for medical analytic mechanisms, scalability of the system, performance issues, network constraints, huge data handling, data security, intelligence in inference, real-time processing, low power consumption and CPU speed management. Syed et al. [8] presented the proposed system in similar ways such as collecting data from multiple wearable sensors placed on the patient's left ankle, right arm, and chest, then transmitting data through MIoT devices to the integrated cloud and data analytics layer. However, to process this huge amount of data, they preferred big data analytics, HadoopMapReduce techniques.

Thus, they achieved 97.1% accuracy of the classification model to predict among 12 physical activities of the person concerned and under observation.

Thus, the MIoT-based system reaches patients without them having to be physically present. This requirement of treating patients has paved the path for using technologies such as radio-frequency identification (RFID) and sensor network interfaced with medical devices where database and communication systems are interfaced to help users such as healthcare systems and medical practitioners. The health system is required to collect huge amounts of data, store the data and analyze the data in real time or when required. This type of infrastructure can be fulfilled with an MIoT-based system. A wireless sensor etwork (WSN) with WSN communication, WSN middleware and proper data security is being employed in MIoT-based healthcare facilities [9,10]. For example, a smartwatch or a body-wearing system can have a temperature sensor to measure body temperature, an oxymeter for measuring oxygen saturation, an accelerometer for monitoring body movement and a GPS to track the position of the patient. Thus, sensors in a wearable device can monitor ECG (electrocardiography) signals, EMG (electromyography) activity and blood pressure. After collection, these data are sent to a cloud-based system and stored there. Sometimes, the data are also analyzed to generate inferences [11]. So, MIoT devices start from wearable things to remote patient monitoring equipment, infusion pumps, sensor-enabled beds and general health monitoring devices. The goal of MIoT is to save patients from the onslaught of the COVID-19 pandemic [12].

Application of machine learning (ML) or deep learning (DL) technologies along with data mining techniques not only reduces computing costs and time but also produces an effective and efficient system that can provide doctors and other health workers with a clear-cut advantage. In the case of heart diseases, ML/DL can provide a very quick analytical report to the healthcare staff/doctors so that next steps of action can be taken very quickly and efficiently [12,13]. Khan and Algarni [12] proposed a concept of using modified salp swarm optimization (MSSO) and an adaptive neuro-fuzzy inference system (ANFIS) to provide prediction capacity in heart disease diagnosis. They employed the Levy flight algorithm in the proposed MSSO-ANFIS to improve the searching capability with an expectation of 99.45% accuracy. Liu and Kim in [13] proposed a classification technique of various heart diseases based on ECG by implementing an ML method called long short-term memory (LSTM). Korurek and Dogan [14] described ECG classification using radial basis function neural network (RBFNN). They also applied particle swarm optimization (PSO). Before that, they filtered the ECG signal and eliminated the base line drift. A low pass linear phase filter was applied to remove noise. After the noise elimination and correction of the base line, they employed the Pan-Tompkins algorithm (Pan & Tompkins, 1985 [15]) to detect R peaks of ECG beats. Their work was improvised by Saini et al. [16] by applying the k-nearest neighborhood (KNN) classification function based on the wavelet features. They chose the KNN classifier for the classification of diseases due to the lower computational burden. Apart from this, feature extraction was efficiently done using the db4 and db5 wavelet for ECG wave form of the patient. There are some other notable works, like Al-Makhadmeh and A. Tolba [17], where the authors illustrate an IoT-based heart disease identification system mainly comprising a higher order Boltzmann deep belief neural network (HOBDBNN). But before the data were received from the IoT and cloud-based system, they were first preprocessed by Median studentized residual approach and then different features were extracted. The authors stressed that a prediction accuracy of 99.03% was possible, which caters to minimizing mortality due to heart disease. In search of accurate arrhythmia detection, Park et al. [18] presented a novel heartbeat classification system that employed an adaptive feature extraction and cascade classifiers for arrhythmia detection, which can be done on smart phone. The proposed method collects ECG recordings of a few continuous hours to a few days and relies on a real-time ECG recognition system working on a smartphone. For pre-processing, Park et al. applied a low-order polynomial. For this, coefficients were chosen matched with the measured ECG signals. Heart beat segmentation was done next based on R-peaks. The random forest cascading method was used for classification with a 97.89% positive predictive value.

Most of the proposed methods of different articles mentioned or in general are based on using ECG signal analysis and feature extraction, but echocardiography is the most common imaging mechanism in cardiovascular diagnoses, which uses ultrasound technology to capture high temporal and spatial resolution images of the heart and surrounding structures. Recent advances in the research of heart disease detection methodologies are using deep learning based convolution neural network (CNN/DCNN), which analyzes the images collected over the internet through an MIoT-based echocardiography device and processes them in a remote location with a health facility [19,20].

10.2 Brief History of MIoT and Its Services

In 1999, Kevin Ashton invented the word "Internet of Things" during his work at Procter & Gamble. Ashton, who was working to improve the supply chain, decided to draw the attention of senior management to a new, innovative technology called RFID. IoT is a structure consisting of interconnected devices or elements inluding electronic hardware, software and sensors. Communication and data transmission are accomplished over the internet among the networked elements using network connectivity and any other devices required for data collection and compilation [16]. The emergence of an IoT-based framework and facilities started penetrating in remote or not easily accessible zones to collect data using devices such as sensors and cameras and also to transmit huge amounts of data using the internet to different zones for compilation and analysis [21]. Manufacture [22–25], agriculture [26], energy [27,28] and GIS (Geographical Information System) [29] sectors are directly governed by this infrastructure. The implementation of IoT-based technologies and procedures in medicine and healthcare also has the ability to minimize costs and increase the quality of healthcare, while at the same time enhancing customer-related services and responses. The MIoT can be described as an IoT network of medical devices, sensors and applications that use computer networks to connect cyber healthcare and physical resources [21].

10.3 Major Challenges in Cardiac Treatment in the Pandemic Period

The lockdown period and also the partial lockdown period have created a worrying situation for CVD patients as their treatment in the majority of cases cannot be stopped. Apart from this, these patients in many cases cannot move easily to reach the diagnosis or healthcare center. They require ambulances, which are not readily available during this time. Moreover, the patients have to come in contact with the healthcare persons or doctors in each case, which is highly risky during the COVID-19 pandemic as they could get infected. In addition, sometimes patients may require instant instruction upon some deterioration so that a heart attack related incident can be avoided. In another situation, a cardiac patient may worsen while traveling by ambulance to the hospital; in that case, finding a nearer or better facility depending upon his or her present clinical status is quite difficult.

10.4 Role of MIoT in a Cardiac Patient's Diagnosis and Treatment

The risk of heart disease can be minimized with early diagnosis and treatment. One solution to the COVID-19 situation for CVD patients is to use of the MIoT-based system to provide quick or instant diagnosis and treatment with no or very low risk of exposure to the outside world. Figure 10.1 represents the role of MIoT in cardiology in a cyclic manner.

The MIoT devices collect data from the cardiac patient before or after the patient develops any complication. Different heart-related or health-related data can be collected remotely when required or in real-time. The collected data will be stored in the remote data server or in the cloud. Thus, the system is efficient and increases the reachability and cost effectiveness of the healthcare system. In this system, the patient benefits the most [31]. The concerned hospital will fetch data from the data server. The data can consists of numeric values (e.g., blood pressure data, oxygen saturation, pulse rate) or image data (e.g., ECG curves, echocardium images). The patient's data should be classified and analyzed by ML/DL-driven application at the hospital site. As a result, a quick inference can be done to help doctors take necessary actions promptly and easily. Figure 10.2 is a compact representation of the MIoT-based system.

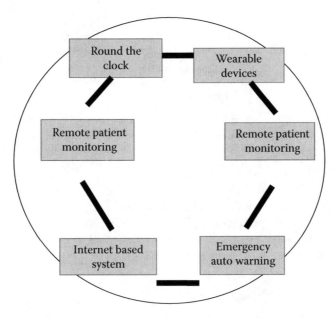

FIGURE 10.1
Role of MIoT in cardiology.

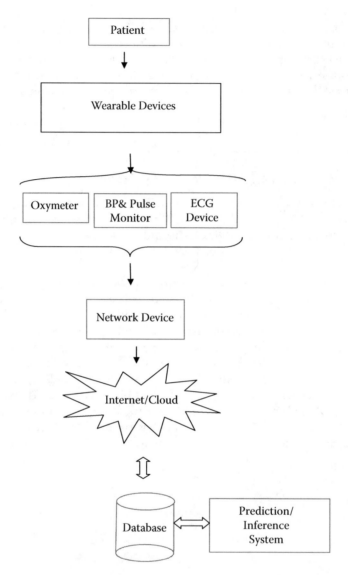

FIGURE 10.2
A compact representation of the MIoT-based system.

A smart watch is a wearable devices with different sensors to record a patient's clinical data such as oxymetry, blood pressure, ECG, etc. Interconnectivity is used to send the data to either the cloud, which is the hospital's own infrastructure, or sometimes stored in a server of the hospital locally. The last module of this system is the prediction system. This is an intelligent part of the system that analyzes data and then infers the situation of the patient.

10.4.1 Heart Disease Prediction System

The heart condition of the patient is monitored and identified remotely after the data are stored in a data center. Usually the proposed systems are performed on training- and testing-based methodologies (i.e., supervised machine learning system). For training of the system, a standard data repository, such as the UCI data repository, is used [32,33]. First, the data values from the dataset undergo preprocessing. Then, the patient's data are tested by comparing and classification using the training result [12,33]. For this supervised learning-based system, some common steps are as follows:

i. Preprocessing

After the data values are collected from the training dataset (e.g., UCI dataset), they are preprocessed to remove noise or to generate missing data. For example, the base line wander, motion artefacts and other interruptions of the original recorded signal are removed [34]. This noiseless corrected data becomes useful to finding patterns associated with heart disease. The patterns of curves for each disease are stored as knowledge during the training phase. These noiseless patterns become capable of recognizing heart disease. Different methods such as the median studentized residual approach [12], symbolic aggregate approximation (SAX) [35], least mean square (LMS) and delayed error normalized LMS (DENLMS) adaptive filter are very effective algorithms in this respect [36].

ii. Feature selection

The purpose of feature selection is to reduce the size of the input dataset so that when further processing and analysis are done, the most effective part of the information is carved out. This mechanism enhances the performance of the process of prediction. It also improves the modeling process.

Table 10.1 presents some related works where the progress on the MIoT-based smart healthcare system is briefed. Sl. No. 1 uses MIoT as well as cloud infrastructure for collection and transmission/storing of patients' health data. It uses modified Salp swarm optimization (MSSO) for optimization purposes and an adaptive neuro-fuzzy inference system (ANFIS) for detection of disease from the data collected and achieves a 99.45% accuracy rate. Sl. No. 2 can be used for an MIoT-based system as well as a cloud environment. The data collection can be implemented using MIoT devices with some improvement over the proposed system. The system is well designed and provides 98.4% accuracy in the classification process with less response time. Sl. No. 3 is a proposed system working on an MIoT-based infrastructure. This work used a higher order Boltzmann deep belief neural network and got a 99.03% accuracy rate in analyzing heart disease recognition. SI No. 4 uses wearable ECG sensors to collect heart data of a patient and transmits the data using Bluetooth or similar devices. It acquires an accuracy rate of 97.34%, which is comparatively better than some previous works. SI No. 5 represents a proposed system that uses sensors for collecting data on the patient and only allows entering the respiratory rate manually by the user in a phone-based application. Later, the data are classified by a rule-based classifier. SI No. 6 has the potential to be implemented in an MIoT-based system because ECG data needs to be collected. A smartphone-based application is a better option for that purpose. A 96% classification accuracy is achieved by this proposed SVM-based classification system. Authors have compared with other techniques like ANN & PCA, KNN classifier, Fuzzy decision classifier, Fuzzy KNN and Neuro-fuzzy but the proposed system showed a much better accuracy rate. Lamonaca et al. [37], or SI No. 7, present the use of MIoT for blood pressure (BP) monitoring. The paper focuses on the structure of MIoT for monitoring BP. It also discusses the need for sensors for that purpose. The authors present their model in five modules. They emphasize the photopletismographic (PPG) signal for recording BP signals. Then, they proces and

TABLE 10.1

Brief Analysis of Some Related Works

Sl. No.	Ref. No.	Use of MIoT	Framework	Methodology
1	12	Yes	MIoT and cloud environment for prediction of heart disease	Modified Salp swarm optimization (MSSO) with ANFIS
2	13	Potential	Preprocessing and classification of ECG signals are proposed	Preprocessing—Symbolic aggregate approximation (SAX). Classification—Long short term memory networks
3	16	Yes	MIoT-based system with prediction of heart disease	Higher order Boltzmann deep belief neural network (HOBDBNN)
4	17	Yes	MIoT/sensor, IoT-based heart monitoring system	Preprocessing—Low-order polynomial with coefficients matched to the measured ECG signals. Classification—Random forest-based cascade
5	29	Yes	MIoT, smartphone-based heart defect monitoring system	Rule-based classifier used to classify vital signs of children
6	35	Potential	ECG signal processing for remote healthcare systems	Preprocessing—Delayed LMS adaptive filter. Feature extraction—Discrete wavelet transform (DWT) function. Segmentation & Classification—Support vector machine (SVM)
7	36	Yes	MIoT for BP recording, processing, elaborating and transmission of readings	Photopletismographic (PPG) analysis, analog to digital conversion (ADC)
8	37	Yes	MIoT-based smart T-shirt for wearable devices	Wearable devices, intelligent devices for communication and a Decision Support System
9	38	Potential	Information security for MIoT, sensor-based health monitoring system.	Multilayer perception model is the fundamental logic used. Signal remains unaffected by interference due to a secure communication channel.
10	39	Yes	MIoT and deep learning in Edge computing devices	Preprocessing—Principal component analysis (PCA), deep learning and neural network

elaborate the acquired signals so that the user can understand the BP signals in terms of values only. Lastly, in the data distribution service module, the BP values are stored and the system presents the stored data using a graphical user interface (GUI). In SI No. 8 [38], Balestrieri et al. designed an MIoT-based system by introducing T-shirts with wearable devices embedded. Their whole system consists of (a) a smart wearable device (smart shirt) that is embedded with sensors for monitoring ECG, measuring respiration rate, measuring skin temperature, measuring galvanic skin response and monitoring human activity; (b) an intelligent communicating device (S-box) that communicates domestically with the smart T-shirt and a remote server via WWAN (wireless wide area network), (c) a decision support system (DSS) that is connected with the S-Box for patient data, (d) a monitoring station that interacts with the DSS via the internet, monitors the patient's status and generates messages or alarms according to instructions from the DSS. The authors have proposed architectures for the smart T-shirt (S-WEAR) containing sensors to tap the patient heart-related data, S-Box. In SI No. 9 [39], Pirbhulal et al. present an ML-based biometric security system that is applicable for MIoT-based healthcare (e.g., heart monitoring) system. Since there is a high requirement of securing health-related data, this paper focuses on the issue with a comprehensive study. The authors have used the ECG signal to cater to the training and testing phases of the proposed system for entity identification. The key mechanism based on the direct sequence spread spectrum (DSSS) method is used for random and unique keys. The ML-based training system is introduced to track an authorized or unauthorized user easily. The last reference, SI No. 10 [40], gives a solution to the limitation in scalability of cloud technology, and as a consequence, the MIoT infrastructure gets less service in the case of IoT and cloud-based automatic heart diagnosis. Shreshth Tuli et al. utilize fog and edge computing to provide low latency and energy-efficient mechanisms for data processing. They use deep learning and IoT. In this model, IoT/MIoT devices collect heart-related data at the patient's location and then send those data, preferably through a smartphone-based app, to a FogBus framework. This framework consists of a data storage facility, a knowledge base system, a cloud integrator subsystem and a deep learning model for classification of data that leads to prediction of disease or patient status. Apart from this, the PCA method is employed for preprocessing of the raw signal. A comprehensive topology of the proposed system is also suggested. Application of deep learning and FogBus has made this proposed system highly efficient over many previous systems.

The research works show that the development of sensors and wearable devices as well as data processing and prediction mechanisms has made good progress. The data collected using IoTs and measuring devices are sent to the cloud and stored for remote or local access when required. The stored data are fetched and preprocessed, then analyzed using ML/DL for prediction or classification purposes. Hence, the system serves as a huge support to the healthcare system or doctor in treating a patient remotely and quickly.

10.5 Requirement of Security Features

In the context of the MIoT network, the following security objectives are usually identified [41]:

A. Confidentiality

It guarantees that secret information is not leaked or made available to unauthorized individuals [42]. When an MIoT-based healthcare system is concerned, the word "confidentiality" refers to secrecy or privacy of patient information shared with healthcare staff, technicians or doctors. Disclosure of these personal data to unauthorized persons can harm the patient concerned in many ways [43].

B. Integrity

This feature assures that the patient data collected are maintained properly and those are neither destroyed nor tampered with to any extent [42]. In the case of MIoT, integrity means keeping patient data accurate and intact [42,43]. Since MIoT devices typically run in less confidential environments, they are often subject to physical attacks that target devices [44].

C. Authenticity

Both the entity authentication and message authentication are involved in this issue. Entity authentication of identification is the mechanism by which the asserted identity of another individual participating in the interaction is guaranteed to one interacting party, and the latter has actually participated. On the other hand, authentication of message is the mechanism by which an entity is checked as an original source of data produced in the past [42]. Since many IoT devices do not have enough memory and CPU capacity to perform a cryptographic operation in this regard, lightweight authentication mechanism are being provided [45].

D. Authorization

Authorization is a security mechanism to determine access levels or privileges of users related to system resources [42]. For example, only trusted expertise parties are allowed to carry out a given service such as issuing commands to MIoT devices or upgrading MIoT device software [45].

E. Availability

The feature ensures that the system works as per requirement and the services are readily available to the users [46]. For MIoT, the required devices and network resources should be available to the patient as per requirement.

F. Non-repudiation

By this, any previous action, commitment or transaction cannot be denied by the entity that initiated it [42]. Items sent from the patient side to the MIoT-based system cannot be denied by the patient later. Likewise, doctors cannot modify the patient's data by denying the original data came from the patient side. This way, disputes related to denial of action can be prevented. Sometimes, a specific mechanism involving a trusted third party is used to a resolve dispute, if any [42].

10.6 Conclusion

An MIoT-based heart monitoring system is very useful, not only for patients located in remote areas but also during COVID-19 pandemic situations. Many times, a cardiac patient is vulnerable to quick or difficult movements and also their condition can deteriorate abruptly, so this facility is suitable for getting treated or diagnosed remotely. But there are still huge scopes of research work on wearable devices and prediction/DSS. Less costly sensors and wearable devices will make the system more popular. Once this system is widely used by common people, the number of heart patients as well as the death rate due to heart disease will be reduced drastically. More improved prediction or DSS for this system will effectively cater to the healthcare staff and doctors; hence, patients will experience gains.

References

1. J. Beard, R. Suzman (2018), "Global health and aging", (online available at:) www.who.int/ageing/publications/global_health.pdf, [Online; accessed 20-March-2018].

2. V. Chaturvedi, N. Parakh, S. Seth, B. Bhargava, S. Ramakrishnan, A. Roy, A. Saxena, N. Gupta, P. Misra, S. K. Rai, K. Anand, C. S. Pandav, R. Sharma, S. Prasad (2016), "Heart failure in India: The INDUS (INDiaUkieri Study) study", *Journal of the Practice of Cardiovascular Science*, vol. 2, no. 1, Page: 28–35, doi: 10.4103/2395-5414.182988

3. Global Burden of Disease Report, 2017, (available online at:) http://www.indiaspend.com/indias-great-challenge-health-sanitation/heart-disease-kills-1-7-million-indianscontinues-to-be-top-killer-in-2016.

4. K. Bhagchandani, D. Peter Augustine (2019), "IoT based heart monitoring and alerting system with cloud computing and managing the traffic for an ambulance in India", *International Journal of Electrical and Computer Engineering (IJECE)*, vol. 9, no. 6, pp. 5068–5074 ISSN: 2088-8708, doi: 10.11591/ijece.v9i6.pp5068-5074.

5. B. Lin, S. Wu (2020), "COVID-19 (Coronavirus Disease 2019): Opportunities and challenges for digital health and the Internet of Medical Things in China", *OMICS: A Journal of Integrative Biology*, vol. 24, no. 5, pp. 231–232.

6. R. P. Singh, M. Javaid, A. Haleem, R. Vaishya, S. Ali (2020), "Internet of Medical Things (IoMT) for orthopaedic in COVID-19 pandemic: Roles, challenges, and applications", *Journal of Clinical Orthopaedics Trauma*, 11(4), pp.713–717. 10.1016/j.jcot.2020.05.011

7. G. J. Joyia, R. M. Liaqat, A. Farooq, S. Rehman (2017), "Internet of Medical Things (IOMT): Applications, benefits and future challenges in healthcare domain", *Journal of Communication*, 12(4), pp. 240–247.

8. L. Syed, S. Jabeen, , A. Alsaeedi (2019), "Smart healthcare framework for ambient assisted living using IoMT and big data analytics techniques", *Future Generation Computer Systems*, vol. 101, pp. 136–151, 10.1016/j.future.2019.06.004

9. J. Gubbi, R. Buyya, S. Marusic, M. Palaniswami (2013), "Internet of Things (IoT): A vision, architectural elements, and future directions", *Future Generation Computer System*, vol. 29(7), pp. 1645–1660. 10.1016/j.future.2013.01.010

10. S. Rani, S. H. Ahmed, R. Talwar, J. Malhotra (2017), "IoMT Song H. A reliable cross layer protocol for internet of multimedia things", *IEEE Internet Things Journal*, 4(3), pp. 832–839.

11. Z. Ning et al., "Mobile Edge Computing Enabled 5G Health Monitoring for Internet of Medical Things: A Decentralized Game Theoretic Approach", in *IEEE Journal on Selected Areas in Communications*, vol. 39 no. 2, pp. 463–478, Feb 2021, 10.1109/JSAC.2020.3020645.

12. M. A. Khan, F. Algarni (2020), "A healthcare monitoring system for the diagnosis of heart disease in the IoMT cloud environment using MSSO-ANFIS", in *IEEE Access*, vol. 8, pp. 122259–122269, doi: 10.1109/ACCESS.2020.3006424

13. M. Liu, Y. Kim (2018), "Classification of heart diseases based on ECG signals using long short-term memory", *40th Annual International Conference of the IEEE Engineering in Medicine and Biology Society (EMBC)*, Honolulu, HI, pp. 2707–2710. doi: 10.1109/EMBC.2018.8512761

14. M. Korurek, B. Dogan (2010), "ECG beat classification using particle swarm optimization and radial basis function neural network," *Expert Systems with Applications*, vol. 37, no. 12, pp. 7563–7569.

15. J. Pan and W. J. Tompkins, "A Real-Time QRS Detection Algorithm," in *IEEE Transactions on Biomedical Engineering*, vol. BME-32, no. 3, pp. 230–236, March 1985, doi: 10.1109/TBME.1985.325532.

16. R. Saini, N. Bindal, P. Bansal (2015), "Classification of heart diseases from ECG signals using wavelet transform and kNN classifier", *International Conference on Computing, Communication & Automation*, Noida, India, pp. 1208–1215, doi: 10.1109/CCAA.2015.7148561.

17. Z. Al-Makhadmeh, A. Tolba (2019), "Utilizing IoT wearable medical device for heart disease prediction using higher order Boltzmann model: A classification approach," *Measurement*, vol. 147, pp. 106815, DOI: 10.1016/j.measurement.2019.07.043

18. J. Park, M. Kang, J. Gao et al. (2017), "Cascade classification with adaptive feature extraction for arrhythmia detection", *Journal Medical System*, 41, 11 10.1007/s10916-016-0660-9.

19. G. Litjens, F. Ciompi, J. M. Wolterink, B.de Vos, T. Leiner, J. Teuwen, I. Išgum, (2019), "State-of-the-art deep learning in cardiovascular image analysis", *JACC Cardiovasc Imaging*, 12(8 Pt 1):1549–1565. 10.1016/j.jcmg.2019.06.009

20. A. Ghorbani, D. Ouyang, A. Abidet al., (2020), "Deep learning interpretation of echo-cardiograms.npj Digit", *Medicine* 3, 10. 10.1038/s41746-019-0216-8.

21. J. Cecil, A. Gupta, M. Pirela-Cruz, P. Ramanathan, (2018), "An IoMT-based cyber training framework for orthopedic surgery using next generation internet technologies". *Informatics in Medicine Unlocked*, vol. 12, pp. 128–137 ISSN 2352-9148, https://doi.org/10.1016/j.imu.2018.05.002

22. H. G., Bader, A. Kadri, M. S. Alouini (2016), "Front-end intelligence for large-scale application-oriented internet-of things", *IEEE Access*, vol. 4, pp. 3257–3272.

23. C. Yang, W. Shen, X. Wang (2016), "Applications of Internet of Things in manufacturing", *2016-IEEE 20th International Conference on Computer Supported Cooperative Work in Design (CSCWD)*, Nanchang, pp. 670–675.

24. M. Bozzi et al. (2016), "Novel materials and fabrication technologies for SIW components for the Internet of Things", *2016-IEEE International Workshop on Electromagnetics: Applications and Student Innovation Competition (iWEM)*, Nanjing, pp. 1–3.

25. Z. Qiu et al. (2016), "IoTI: Internet of things instruments reconstruction model design", *2016 IEEE International Instrumentation and Measurement Technology Conference Proceedings*, Taipei, pp. 1–6.

26. M. Lee, J. Hwang, H. Yoe (2013), "Agricultural production system based on IoT", *IEEE 16th Int. Conf. Comput. Sci. Eng.*, pp. 833–837.

27. B. Martinez, M. Monton, I. Vilajosana, J. D. Prades (2015), "The power of models: Modeling power consumption for IoT devices," *IEEE Sensors Journal*, vol. 15, no. 10, pp. 5777–5789.

28. J. Pan, R. Jain, S. Paul, T. Vu, A. Saifullah, M. Sha (2015), "An Internet of Things framework for smart energy in buildings: Designs, prototype, and experiments", *Internet of Things Journal, IEEE*, 2(6), 527–537.

29. S. Liu, G. Zhu (2014), "The application of GIS and IOT technology on building fire evacuation", *Procedia Engineering*, vol. 71, no. 2014, pp. 577–582, 10.1016/j.proeng.2014.04.082

30. A. Fatmah et al. (2017), "Congenital Heart Defect Monitoring System for Children using Internet of Things", *International Journal of Computer Applications (0975 – 8887)*, vol. 171, no. 3, pp. 44–48.

31. G. J. Joyia, R. M. Liaqat, A. Farooq, S. Rehman (2017), "Internet of Medical Things (IOMT): Applications, benefits and future challenges in healthcare domain", *Journal of Communications*, vol. 12, no. 4, pp. 240–247, doi: 10.12720/jcm.12.4.240-247.

32. Kaggle open dataset, URL: https://www.kaggle.com/datasets, accessed on Apr 15, 2020, 3:30A.M. (https://www.kaggle.com).

33. K. Uyar, A. İlhan (2017), "Diagnosis of heart disease using genetic algorithm based trained ecurrent fuzzy neural networks", *Procedia Computer Science*, vol. 120, no. 2017, pp. 588–593, 10.1016/j.procs.2017.11.283.

34. H. M. Rai, A. Trivedi, S. Shukla (2013), "ECG signal processing for abnormalities detection using multiresolution wavelet transform and Artificial Neural Network classifier," *Journal of Measurement*, vol. 46, no. 9, pp. 3238–3246.

35. J. Lin, E. Keogh, S. Lonardi, B. Chiu (2003), "A symbolic representation of time series, with implications for streaming algorithms," in *Proceedings of the 8th ACM SIGMOD workshop on Research issues in data mining and knowledge discovery*, pp. 2–11.

36. C. Venkatesan, P. Karthigaikumar, A. Paul, S. Satheeskumaran, R. Kumar (2018), "ECG signal pre-processing and SVM classifier-based abnormality detection in remote healthcare applications", in *IEEE Access*, vol. 6, no. 2018, pp. 9767–9773, doi: 10.1109/ACCESS.2018.2 794346.
37. F. Lamonaca et al. (2019), "An overview on Internet of Medical Things in blood pressure monitoring", *2019 IEEE International Symposium on Medical Measurements and Applications (MeMeA)*, Istanbul, Turkey, pp. 1–6, doi: 10.1109/MeMeA.2019.8802164.
38. E. Balestrieri et al. (2019), "The architecture of an innovative smart T-shirt based on the Internet of Medical Things paradigm", *IEEE International Symposium on Medical Measurements and Applications (MeMeA)*, Istanbul, Turkey, pp. 1–6, doi: 10.1109/MeMeA.2019.8802143.
39. S. Pirbhulal, N. Pombo, V. Felizardo, N. Garcia, A. H. Sodhro, S. C. Mukhopadhyay (2019), "Towards machine learning enabled security framework for IoT-based healthcare", *13th International Conference on Sensing Technology (ICST)*, Sydney, Australia, pp. 1–6, doi: 10.11 09/ICST46873.2019.9047745.
40. S. Tuli, N. Basumatary, S. S. Gill, M. Kahani, R. C. Arya, G. S. Wander, R. Buyya (2019), "Health Fog: An ensemble deep learning based Smart Healthcare System for Automatic Diagnosis of Heart Diseases in integrated IoT and fog computing environments", *Future Generation Computer System*, 10.1016/j.future.2019.10.043.
41. P. Maria et al. (2020), "A survey on security threats and counter measures in Internet of Medical Things (IoMT)", *Transaction on Emerging Telecommunications Technologies, Wiley*, doi: 10.1002/ett.4049.
42. J. Menezes Alfred, C. Oorschot Paul, S. A. Vanstone (1996), *"Handbook of applied cryptography"*, CRC Press, ISBN: 0-8493-8523-7 1996.
43. H. Joan, B. Pauline, J. Arnold, et al. (2008), *"An Introductory Resource Guide for Implementing the Health Insurance Portability and Accountability Act (HIPAA) Security Rule Technology Administration"*, Report – Illinois: American Health Information Management Association.
44. I. Makhdoom, M. Abolhasan, J. Lipman, R. P. Liu, W. Ni (2019), "Anatomy of threats to the Internet of Things", *IEEE Communication Survey Tutor*, vol. 21(2), pp. 1636–1675, 10.1109/ COMST.2018.2874978.
45. A. Alrawais, A. Alhothaily, C. Hu, X. Cheng (2017), "Fog computing for the Internet of Things: Security and privacy issues", *IEEE Transaction on Internet Computer*, vol. 21(2), pp. 34–42, 10.1109/MIC.2017.37.
46. D. Jyoti, V. Amarsinh (2017), "Security attacks in IoT: A survey", *International Conference on I-SMAC (IoT in Social, Mobile, Analytics and Cloud) (I-SMAC)*, India; pp. 32–37.

11

MIoT: Paralyzed Patient Healthcare

Kinjal Raykarmakar, Shruti Harrison and Anirban Das
*University of Engineering and Management,
Kolkata, India*

CONTENTS

11.1 Introduction

The paralyzed patient's glove is a device that can assist patients with paralysis due to spinal cord injury or illness. Analysis of the glove movement made by the patient helps to clarify the needs, emotions and thoughts that the patient wants to share or perform. Such devices will be helpful for the rehabilitation of the patient. The following section elaborates on issues surrounding the sensor-based glove.

11.1.1 Paralyzed Patient's Glove and Its Importance

Among the many advancements in the medical sector, very few focus on helping patients with disabilities to communicate. Although monitoring systems make it easier for doctors to collect and observe a patient's vitals, there are not many options to help patients with

DOI: 10.1201/9780429318078-11

disabilities communicate verbally. We propose a simple yet effective way to solve this age-old problem. The main purpose is to replace the conventional approach of patient–nurse communication with a modern technology that provides a much faster and reliable way. In the current scenario, the patient has to be dependent on a family member or a nurse, both of whom have to attend to the patient constantly. Our objective is to allow such patients to communicate independently with the nurse by tilting a device located on his or her finger or another part of the body that is capable of movement.

This will not only help the patient but also will ease the nurse's or caregiver's burden. In a hospital, because a single nurse is responsible for a number of patients, the nurse will save time if he or she does not have to visit every patient to meet his or her needs. This device can be implemented at home, too. After the patient sends a message, the nurse, family member or caregiver can remotely monitor the request and provide assistance without any further delay. To make the system more dynamic, the following ideas focus on building a smart system to make the patient self-sufficient as well as to assist nurses, doctors and family members.

11.1.2 Important Modules and Sensors

MPU 6050 [1]

- Input voltage: 2.3–3.4 V
- Tri-axis angular rate sensor (gyro) with a sensitivity up to 131 LSBs/dps and a full-scale range of ±250, ±500, ±1,000, and ±2,000 dps
- Tri-Axis accelerometer with a programmable full-scale range of ±2 g, ±4 g, ±8 g and ±16 g

11.1.3 Working Principle

The MPU 6050 is a sensor based on integrated 3-axis MEMS (micro electro mechanical systems) technology containing an accelerometer and a gyroscope. Both the accelerometer and the gyroscope are embedded inside a single chip. The MPU 6050 is a 6 DOF (degree of freedom) or a 6-axis IMU (inertial measurement unit) sensor; that is, it will give 6 values in output, three values from the accelerometer, and three from the gyroscope. This chip uses an I²C (inter-integrated circuit) protocol for communication [2].

- An accelerometer works on the principle of the Piezoelectric effect. Whenever the sensor is tilted, the ball will move in that direction because of gravitational force. The walls are made of Piezoelectric elements, so every time the ball touches the wall an electric current will be produced, which will be interpreted in the form of values in any 3D space [2].
- A gyroscope is a spinning wheel or disc in which the axis of rotation is free to assume any orientation by itself. When rotating, the orientation of this axis is unaffected by tilting or rotation of the mounting. Because of this, gyroscopes are useful for measuring or maintaining orientation [2].

The system makes use of a microcontroller-based circuitry to achieve this functionality. It makes use of a hand motion recognition circuit and a receiver plus transmitter circuit. The hand motion circuit is used to detect hand movements using the accelerometer and the gyroscope and then transmit this information wirelessly over radio frequency (RF) to the receiver system. The receiver system is designed to receive and process these commands and display them over the LCD display as well as transmit the data online over to an Internet of Things (IoT) Gecko server. The IoT Gecko server then displays this information online, to achieve the desired output.

Heart rate sensor (SEN-11574) [3]:

Kit includes:

- Pulse sensor board
- Velcro finger strap

Heart rate data can be useful for designing an exercise routine, studying our activity or anxiety levels with our heart beat. The problem is that heart rate can be difficult to measure but the pulse sensor amped can solve this problem. The pulse sensor amped is a plug-and-play heart rate sensor for the microcontroller.

Temperature sensor (LM35) [4]:

LM35 is a precession integrated circuit temperature sensor whose output voltage varies based on the temperature around it. It is a small, cheap integrated circuit that can be used to measure temperature anywhere between −55°C and 150°C. It can easily be interfaced with any microcontroller that has Analog to Digital Converters function or any development platform such as Arduino.

If the temperature is 0°C, then the output voltage will also be 0 V. There will be a rise of 0.01 V (10 mV) for every degree Celsius rise in temperature.

11.1.4 Block Diagrams

Sender (Glove):

An RF transmitter uses different components to transmit position data, temperature data, and even current program register values wirelessly to the receiver. These modules have a range of up to 500 ft in open space. The transmitter operates from 2 V to 12 V. The higher the voltage, the greater the range. We have used these modules extensively and have been very impressed with their ease of use and direct interface to an MCU. The theory of operation is very simple. What the transmitter "sees" on its data pin is what the receiver outputs on its data pin. If you can configure the UART module on auC, you have an instant wireless data connection. The typical range is 500 ft for an open area. This is an ASK transmitter module with an output of up to 8 mW depending on the power supply voltage. The transmitter is based on SAW resonator and accepts digital inputs, can operate from 2 V-DC to 12V-DC, and makes building RF-enabled products very easy [5] (Figure 11.1).

Receiver:

The receiver type is good for data rates up to 4,800 bps and will only work with the 434 MHz or 315 MHz transmitter. Multiple 434 MHz or 315 MHz receivers can listen to one 434 MHz transmitter or 315 MHz transmitter. This wireless data is the easiest to use—the lowest cost RF link we have ever seen! Use these components to transmit position data, temperature data and even current program register values wirelessly to the receiver. These modules have a range up to 500 ft in open space. The receiver is operated at 5 V. We have used these modules extensively and have been very impressed with their ease of use and direct interface to an MCU. The theory of operation is very simple. What the transmitter "sees" on its data pin is what the receiver outputs on its data pin. If you

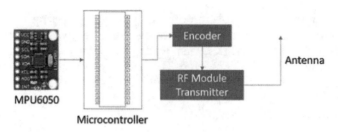

FIGURE 11.1
Block diagram of sender.

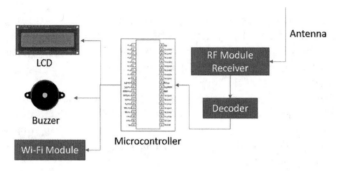

FIGURE 11.2
Block diagram of receiver.

can configure the UART module on auC, you have an instant wireless data connection. Data rates are limited to 4,800 bps. The typical range is 500 ft for open area. This receiver has a sensitivity of 3 uV. It operates from 4.5 V-DC to 5.5 V-DC and has a digital output. The typical sensitivity is −103 dbm, and the typical current consumption is 3.5 mA for 5 V operation voltage [6] (Figure 11.2).

The Glove:

The receiver, mainly the gyroscope and the microcontroller, along with the RF transmitter has to be mounted in a glove, for the user to wear. It is important to correctly place the gyroscope in the glove for proper setup of all the commands. The diagram given shows the proper placement of the gyroscope in the glove. This alignment is in sync with the pseudocode. *Note:* The Y-axis points to the direction of the fingers, and the X-axis points to the left direction. The direction of the rotating arrows is taken as positive values of the angular velocities (Figure 11.3).

11.1.5 Hardware

The purpose of the hardware is to make physical objects responsive and give them the capability to retrieve data from the sender and then process the data and send the results to output devices on the basis of the output and interaction with the software. In our case, the hardware devices are

    i. I/O Devices:
       Glove: It receives the data from the patient and sends it to the processing unit.
       The glove is set up with sensors and modules. The sensors receive the data from

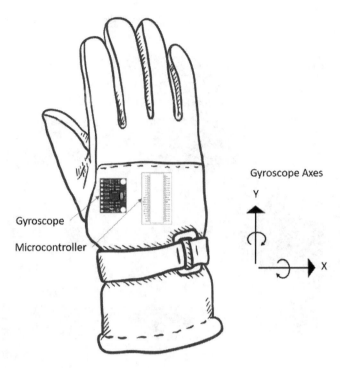

FIGURE 11.3
Block diagram of glove fitted with gyro and microcontroller.

the surroundings and send the data to the microcontroller unit.
LED: The LED views the result after the data is processed.

ii. Processing Unit:
   Microcontroller Unit: Here the data is processed from the raw data, and the unit gives the output to interact with the patient.

iii. Connecting Module:
   The WiFi module is the connecting module which connects the sender and receiver with a high level of accuracy and real-time interaction.

11.1.6 Software

Software must enable data collection, storage, processing, manipulation and instruction. These help in regulation and maintenance of the health of the patient at regular intervals and provide data about whether the patient is improving or deteriorating. Interaction with the patient is the most important thing since the patient has paralysis.

Communication: Most important of all is the communication infrastructure, which consists of protocols and technologies that enable two physical objects to exchange data.

Methodology: In this monitoring and interactive system, the implementation of the module is for paralyzed patients. This module comprises parameters including heart rate, respiration rate and temperature. The primary function of this system is to monitor the heart rate, breathing rate and temperature of the patient, and the data collected by the sensors are sent to the MSP430 launchpad. This launchpad will process the sensed data

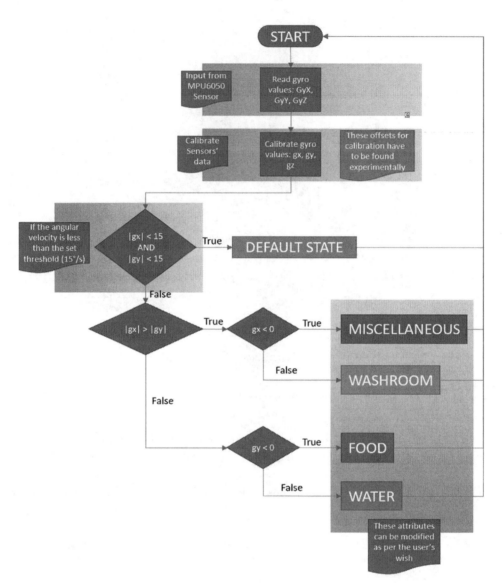

FIGURE 11.4
Flowchart of the entire process.

using the embedded program for the required parameters using code composer studio complier. The program is available for monitoring at operational levels. The normal heart beat range of a paralyzed person will be nearly 60–100 bpm. If the range reaches below 60 bpm, the patient may experience heart block or syncope, and when the range exceeds 100 bpm, the patient may experience anxiety and tachycardia. So, when heart rate increases or decreases, the status of the patient's pulse rate will be detected. The normal temperature of a paralyzed patient is 98.6 F (i.e., 37 C). If the range reaches above or below 98.6 F, the patient can experience irrational thinking and health problems. This can be measured by a change in temperature. The normal respiratory rate of a paralyzed patient can be nearly 12–20 breathes per minute. The rate is usually measured when a person is at rest and involves counting the number of breaths for one minute by counting how many times the chest rises. If the paralysis increases for the patient, then the respiration increases. This can be the basic parameter for paralysis. If the respiration rate increases or decreases 12–20 breathes per minute, it will be detected [7].

All the basic parameters are monitored, and if there is a dangerous change in the patient's status then proper medication can be given at the proper time, and the caregiver will receive the proper information about the patient's condition. This contribution of the paralysis monitoring system is for better process management, superior flexibility and increased efficiency within hospitals, further underlining the appeal of wireless networking options for monitoring systems for paralyzed patients.

Besides monitoring vitals, the glove allows the caregiver to receive proper information about the needs of the patient and to learn about the mental state of the patient because the patient can transmit his or her needs through an electronic signal by moving his or her hand and making gestures with the gyroscope. Proper interaction can be made by the patient to meet his or her needs. Hence, proper caregiving is possible. A flowchart of the entire process is shown in Figure 11.4.

11.1.7 Pseudocode

```
GyX = GyY = GyZ = 0 // (raw data) gyroscope values at three direction
    gx = gy = gz = 0// calibrated values
    function read_mpu6050(){
    /* read the values of gyroscopes from read_mpu6050
    and store them in (GyX, GyY, GyZ) */
    }
    function read_LM35(){
    /*read the values of temperature from LM35
    Store them in t1,t2,.....,tn on hourly basic for 24hrs */
    }
    function display LM35(){
    /* display values from read_LM35()*/
    If normal_temp>avg(t1,t2,...,tn)>normal_temp:
    Print("Medication required")
    Else: Print("Everything is Normal")
    }
    function read_SEN_11574(){
    /*read the values of heart rate from LM35
    Store them in hr1,hr2,......,hrn on hourly basic for 24hrs */
    }
    Function display SEN_11574(){
    /* display the values from read_SEN_11574()
    If normal_hr>avg(hr_1,hr_2,...,hr_n)>normal_hr:
    Print("Medication required")
    Else: Print("Everything is Normal")
    }
```

```
function calibrate_values(){
/* calculate offests via experimentation */
gx = (GyX ± offest) / 131.07
gy = (GyY ± offest) / 131.07
gz = (GyZ ± offest) / 131.07
}
function buzzer(){
/* sound the buzzer*/
}
function default(){
/* send message for "NO INPUT" */
}
function food(){
buzzer()
/* send message for "I NEED FOOD" */
}
function water(){
buzzer()
/* send message for "I NEED WATER" */
}
function washroom(){
buzzer()
/* send message for "I NEED TO GO TO THE WASHROOM" */
}
function miscellaneous(){
buzzer()
/* send message for "I NEED SOMEONE TO ATTEND ME" */
}
while(true){
read_mpu6050()
calibrate_values()
function read_LM35()
display LM35()
read_SEN_11574()
display SEN_11574()
/* abs() return absolute values*/
IF abs(gx) < 15 AND abs(gy) <15{ // if angular velocity is less than 15degrees/sec
/* NO INPUT */
default()
}
ELSE IF abs(gx) > abs(gy){ // x is the major axis of movement
IF gx<0{ // angular velocity over x axis is -ve (anti-clockwise)
/* MOVED DOWN */
miscellaneous()
}
ELSE{ // angular velocity over x axis is +ve (clockwise)
/* MOVED UP */
```

```
washroom()
}
}
ELSE{ // y is the major axis of movement
IF gy<0{ // angular velocity over y axis is -ve (anti-clockwise)
/* MOVED LEFT */
food()
}
ELSE{ // angular velocity over y axis is +ve (clockwise)
/* MOVED RIGHT */
water()
}
}
}
```

11.1.8 Benefits

The IoT-based healthcare system for paralyzed patients will help to monitor the needs of these patients. This system helps the patients overcome barriers to convey their needs without much effort. Moreover, this system can be modified for several purposes where the person's mobility is affected. The IoT-based healthcare system for paralyzed patients is a smart, secure and real-time system that serves as a reliable and efficient system for paralyzed people. The system is inexpensive, is simple and has a dependable assembly. This system also allows patients to ask for help. Some of the advantages of this system are as follows:

- It is highly flexible.
- It has a quick response time.
- It is a real-time application.
- It requires reduced human activity.
- It is a complete patient monitoring system.
- It is highly secure.
- It has high accuracy.

11.1.9 Possible Advancements

This device has made a conveyance of messages possible with only the motion of a body part. By implementing this system, a simple device for people with disabilities can be achieved without the use of complex input.

i. More sensors such as pressure sensors, contact sensors and flexion sensors can be introduced for more accurate and precise data collection. Also, flexion sensors in gloves can be used to decode sign languages created by the hands.
ii. The project can be further developed into an automatic wheelchair wherein the wheelchair will be moved just by hand gesture.

iii. Along with only message transmission, other data such as body temperature, pulse rate, etc., can be transmitted to a nurse and maintained in a real-time record for the patient.

iv. Our system can be used to make gesture-controlled devices such as music apps, cars or smart clocks.

v. Gesture-controlled robotic arms can be designed to grab objects from a distance.

11.2 Challenges in Development

The prototype we have made is fully functional but is restricted to a small area of operation. For a large area and transmission distance, the type of communication used will have to be more effective and faster. For any modular and functional Medical IoT product, there is a need for coordination between hardware and software. Possible challenges might hinder the development and deployment of the products.

11.2.1 Hardware Challenges

i. Power: The design of the product is small and lightweight. Sensors such as gyroscopes will not use much power, but when it comes to the microcontroller, WiFi transreceiver, it draws a lot of power. Also, keep in mind that a glove must be fully autonomous (without any external power supply) to keep it mobile. An internal power supply must be designed. Power distribution, management and optimization throughout the circuit is essential.

ii. Robustness: Sensors and microcontrollers are generally sophisticated in nature. On the other hand, the glove in the hand of a paralyzed person must withstand external pressures. Excessive force on any of the peripherals might lead to damaging them or yielding incorrect sensor values, thus leading to abrupt behavior. The product must also withstand different weather conditions, dust and humidity.

iii. Connectivity: As the whole connectivity used in this product is wireless, it needs to use radio waves. A variety of waves of different frequencies, wavelengths and bandwidths can be used. At the same time, the patient's health should be kept in mind while selecting wave parameters, as these ways might affect the patient's metabolism. Futuristic and safer ways to wireless connectivity such as LiFi can be used, too.

11.2.2 Software Challenges

i. Error handling: Values from sensors might be erroneous, or data might get corrupted during transmission. These errors cannot be avoided and/or immediately fixed. So, software algorithms or multiple substitute sensors can be used to detect and fix these errors.

ii. Debugging: With so many sensors, actuators and real-time data involved, it is difficult to debug an IoT project. So, pseudo inputs under simulated and controlled environments need to be used to test this product.

iii. Computation: Medical IoT devices do not have sufficient memory, computation and communication capabilities; they require a powerful and scalable high-performance computing and massive storage infrastructure for real-time processing and data storage. Hence, there is a need for cloud services.

11.2.3 Network Challenges

i. Internet: One of the biggest challenges in Medical IoT is access to the internet. An IoT device demands an internet connection round the clock. As the micro-processors used have less computing power, data are often transferred to the cloud for fast as well as accurate computation. Also, these computed data need to be sent back for further processing, alerting and/or movement of actuators. So, internet connectivity along with connection speed, stability and reliability, and security are the most important aspects for an MIoT product. As data are processed in real time, there is a need for a continuous and stable internet connection. Downtime might lead to severe consequences. Moreover, research and development in this field, along with the introduction of 5G, can greatly improve this situation.

ii. Protocols and Security: Low-cost devices and software applications based on sensors should follow specific policy and proxy rules to provide services. At present, if we want to provide high-grade security for the sensors, we must apply the high-cost solutions. This is a conflict in Medical IoT systems. Because of the convenience and low cost, a number of devices and software services rely heavily on wireless networks, such as WiFi, which are known to be vulnerable to various intrusions, including unauthorized router access, man-in-the-middle attacks, spoofing, denial of service attacks, brute-force attacks and traffic injections

11.3 Security and Privacy

The security and privacy of patient-related data are two indispensable concepts. By data security, we mean that data are stored and transferred securely, to guarantee their integrity, validity and authenticity, and data privacy means the data can only be accessed by the people who have authorization to view and use them. More reasonable protection strategies could be developed according to different purposes and requirements. In developing medical internet security and privacy systems, the following four requirements should be considered.

i. Data Integrity: Data integrity refers to the fact that all data values satisfy semantic standards without unauthorized tampering. This includes two levels of accuracy and reliability. Data integrity can be divided into four categories, namely, entity integrity, domain integrity, referential integrity and user-defined integrity, which can be maintained by foreign keys, constraints, rules and triggers.

ii. Data Usability: Data usability is to ensure that data or data systems can be used by authorized users. Big data brings not only great benefits but also crucial challenges, such as dirty data and nonstandard data. In addition, data corruption or data loss caused by unauthorized access also further destroys data usability.

iii. Data Auditing: Audit of medical data access is an effective means to monitor the use of resources and a common measure for finding and tracking abnormal events. In addition, cloud service providers usually play untrusted roles, which require reasonable auditing methods. Audit content generally includes users, cloud service providers, access and operation records.

iv. Patient Information Privacy: Patient information can be subdivided into two categories: general records and sensitive data. Sensitive data, which can also be called patient privacy, include mental status, sexual orientation, sexual functioning, infectious diseases, fertility status, drug addiction, genetic information and identity information. We need to make sure that the sensitive data are not leaked to unauthorized users, or even if data are intercepted, the information expressed cannot be understood by unauthorized users.

v. Access Control: Access control is the means by which a data system defines the identity of a user and the predefined policies which prevent access to resources by unauthorized users. There are various encryption methods applied in access control, including symmetric key encryption (SKE), asymmetric key encryption (AKE), and attribute-based encryption (ABE). According to general knowledge, cryptography relies on keys. The size and generation mechanism of secret keys directly affect the security of the cryptosystem. Therefore, for a cryptosystem, key management mechanism determines the security system's life cycle. Owing to the scalable key management and flexible access control policies, ABE is gradually becoming one sort of mainstream method.

11.4 Conclusion

Technology has infiltrated every industry, including healthcare, business, finance and others. Healthcare remains the fastest to adopt technological changes to revolutionize the diagnosis and treatment of the body. IoT offers a multitude of benefits such as improving the effectiveness and quality of services via medical devices. Have a look at a few of the statistics revealing the use of IoT in healthcare and the overall impact on the industry [8]:

- Nearly 60% of healthcare organizations have introduced IoT devices into their facilities.
- 73% of healthcare organizations use IoT for maintenance and monitoring.
- 87% of healthcare organizations plan to implement IoT technology by 2019, which is slightly higher than the current rate of 85% of businesses across various industries.
- Nearly 64% use of IoT in the healthcare industry is patient monitors.
- 89% of healthcare organizations have suffered from IOT-related security breaches.

It is widely known that interconnected devices are being used to gather data from fetal monitors, blood glucose level monitors, electrocardiograms and temperature monitors.

St. John Sepis Agent, DocBox and identification of environmental factors affecting babies in hospitals are three illustrations of the effect that IoT in healthcare has on patients' health and well-being.

References

1. "MPU-6050 Six-Axis (Gyro + Accelerometer) MEMS MotionTracking™ Devices," *TDK InvenSense*, [Online]. Available: https://invensense.tdk.com/products/motion-tracking/6-axis/mpu-6050/. [Accessed 18 September 2020].
2. Module143, "MPU6050 – How to Use it?," 2017. [Online]. Available: http://invent.module143.com/mpu6050-how-to-use-it/. [Accessed 18 September 2020].
3. "Pulse Sensor | sparkfun," [Online]. Available: https://www.sparkfun.com/products/11574. [Accessed 18 September 2020].
4. "LM35 Temperature Sensor," [Online]. Available: https://components101.com/lm35-temperature-sensor. [Accessed 18 September 2020].
5. "RF Link Transmitter – 315MHz | sparkfun," [Online]. Available: https://www.sparkfun.com/products/retired/8945. [Accessed 18 September 2020].
6. "RF Link Transmitter – 434MHz | sparkfun," [Online]. Available: https://www.sparkfun.com/products/retired/8946. [Accessed 18 September 2020].
7. Deepasri T., Gokulpriya M., Arunkumar G., Mohanraj. P. and Shenbagapriya M, "Automated Paralysis Patient Health Care Monitoring System," *South Asian Journal of Engineering and Technology*, vol. 3, no. 2, pp. 85–92, 2017.
8. https://www.smartcity-iot-news.com/, "Internet of Things in Healthcare," 2018. [Online]. Available: https://www.smartcity-iot-news.com/smart-living/internet-of-things-in-healthcare. [Accessed 17 September 2020].

12

Vision–X: IoT-Based Smart Navigation
System for People with Visual Challenges

Tuhin Utsab Paul[1] and Aninda Ghosh[2]
[1]*St. Xavier's University, Kolkata, India*
[2]*Altor, Bangalore, India*

CONTENTS

12.1 Introduction

With the development of the Internet of Things (IoT) infrastructure, it is possible to connect electronically assisted medical equipment to the IoT grid. We implemented an image-based tracking system within IoT for precise locomotive assistance and tracking.

Vision-X is a small electronic device that can be used to guide any person with visual impairments and disabilities within any building, which is known or unknown to the person, and also in outdoor environments such as fields or roads. This is an IoT-based system that connects cameras to a processing unit that analyzes the real-time video, identifies objects and converts the knowledge to audio that is transmitted to earphones worn by the person. The user's locomotion data are sent to the server for creating a digital map of the places visited by the user.

This device aims to facilitate the independent navigation of persons with visual challenges. The device captures real-time video; detects objects, humans and obstructions; and converts the detected information in the form of speech that is transmitted back to the user. By listening to the speech instructions, the user can move around. This device works with very high accuracy in an indoor setup and well as an outdoor setup.

The processor of the device uses different image and video processing algorithms to help the person navigate with ease. For persons with visual impairments, the output will be a computer-generated voice instructing how to navigate, and for persons who are able to see clearly but are not able to move independently, the output will be through the motor controllers, hinged to the wheels of the wheelchair.

This is a navigation system with two webcams mounted on spectacles, an ARM processor development board, simple wireless earphones and a power source. It can be implemented on an Android platform using the processor of the smartphone as it is also an ARM processor, so the user does not need to carry an extra processor with the spectacles. The system can detect any kind of obstacle in its way. It can easily detect doors and staircases, both up and down staircases, as well as human beings. The program generates speech signals as output that are fed to the user via earphones. It is power efficient. When idle, it consumes 1.15 W, and when all cores are being utilized, power consumption goes up to 3 W.

This chapter will offer an understanding of the role of IoT in assisted locomotion of persons with visual impairments in the healthcare sector. A review of studies of various other technologies for assisted locomotion is also provided. The chapter also provides a detailed construct of the Vision-X system and its algorithms. The various areas of application of the IoT-based smart Navigation system for persons with visual impairments is discussed. The chapter is organized as follows: Section 12.1 gives a brief introduction of the chapter. Section 12.2 provides the literature review of the work done in this field. Section 12.3 gives the basic construction of the device and the hardware requirements. Section 12.4 gives a detailed description of the various algorithms that work together to detect objects, humans and obstructions. Section 12.5 gives snapshots of various output scenarios of the trial run that was done. Lastly, Section 12.6 gives the concluding remarks on the device.

12.2 Literature Review

Artificial intelligence has been used to establish techniques for computer vision for supporting blind people and people with visual impairments. Literature disclosed the paradigms for serving people who are unable to travel alone. There are a lot of interesting publications, projects and mobile apps available in this field because this is a burning topic nowadays. Here is an overview of them with significant points.

Volodymyr Ivanchenko in 2008 described Clear Path Guidance to guide blind people and people with visual impairments who use wheelchairs along a clear path that uses computer vision to sense the presence of obstacles or other terrain features and warn the user accordingly. Since multiple terrain features can be distributed anywhere on the ground, and their locations relative to a moving wheelchair are continually changing, it is challenging to communicate this wealth of spatial information in a way that is rapidly comprehensible to the user. To develop a novel user interface that allows the user to interrogate the environment by sweeping a standard (unmodified) white cane back and forth, the system continuously tracks the cane location and sounds an alert if a terrain feature is detected in the direction the cane is pointing. Experiments are described demonstrating the feasibility of the approach [1].

Hend S. AlKhalifa in 2008 proposed Utilizing QR Code. This paper proposed a barcode-based system to help the people with visual impairments and blind people to identify objects in the introduced environment. The system is based on the idea of utilizing QR codes (two-dimensional barcodes) affixed to an object and scanned using a camera phone equipped with QR reader software. The reader decodes the barcode to a URL and directs the phone's browser to fetch an audio file from the web that contains a verbal description of the object. Their proposed system is expected to be useful in real-time interaction with different environments and to further illustrate the potential of this new idea [2].

Jerey P. Bighamy in 2010 introduced VizWiz::LocateIt. This method enables blind people to take a picture and ask for assistance in finding a specific object. The request is first forwarded to remote workers who outline the object, enabling efficient and accurate automatic computer vision to guide users interactively from their existing cell phones. A two-stage algorithm is presented that uses this information to conduct users to the appropriate object interactively from their phone. Researchers produced an app in iOS. The app is available in iTunes, so iPhone users can use it. They are trying to make the app for the Android platform [3].

Jing Su in September 2010 introduced Timbremap, a sonification interface enabling users with visual impairments to explore complex indoor layouts using off-the-shelf touchscreen mobile devices. This is achieved using audio feedback to guide the user's finger on the device's touch interface to convey geometry. The user study evaluation shows Timbremap is effective in conveying nontrivial geometry and enabling users with visual impairments to explore indoor layouts.

Roberto Manduchi in 2010 explored Blind Guidance using mobile computer vision. Researchers focused on the usability of a wayfinding and localization system for persons with visual impairments. This system uses special color markers, placed at key locations in the environment, that can be detected by a regular camera phone. Three blind participants tested the system in various indoor locations and under different system settings. Quantitative performance results are reported for the different system settings. The reduced field of view setting is the most challenging one. The role of frame rate and marker size did not have clear results in the experiments.

Barcodes Ender Tekin and James M. Coughlan in 2010 presented Find and Read Product Barcodes. They described a mobile phone application that guides users with visual impairments to the barcode on a package in real time using the phone's built-in video camera. Once the barcode is located by the system, the user is prompted with audio signals to bring the camera closer to the barcode until it can be interpreted by the camera. Then, it is decoded, and the corresponding product information is read aloud using a text-to-speech converter. Experiments with a blind volunteer demonstrate proof of concept of the system, which allowed the volunteer to locate barcodes that were then translated to product information that was announced to the user.

Y. Tian in 2010 explored Door Detection Indoor Wayfinding. Researchers presented a robust image-based door detection algorithm based on doors' general and stable features (edges and corners) instead of appearance features (color, texture, etc.). A generic geometric door model is built to detect doors by combining edges and corners. Furthermore, additional geometric information is employed to distinguish doors from other objects with similar size and shape (e.g., bookshelf, cabinet, etc.). The robustness and generalizability of the proposed detection algorithm are evaluated against a challenging database of doors collected from a variety of environments over a wide range of colors, textures, occlusions, illuminations, scales, and views.

F. Bellotti in May 2011 presented LodeStar, a location-aware multimedia mobile guide. LodeStar was designed to enhance fruition and enjoyment of cultural and natural heritage for persons with visual impairments. It aimed to support persons with visual impairment in the construction of location awareness by providing them with location-related added value information about object-to-self and object-to-object spatial relations. It presented the underlying psychological and theoretical basis, the description of the CHI design of the mobile guide and user test results coming from trials conducted with real users in two contexts of authentic use: the Galata Sea Museum and the Villa Serra naturalistic park in Genoa. Results mainly revealed that the guide can contribute to people's ability to construct overall awareness of an unfamiliar area, including geographical and cultural aspects. The users still appreciate the guide descriptions of points of interest in terms of content and local/global spatial indications. Interviews highlighted improving the visit experience of persons with visual impairment.

João José in May 2011 invented a system called Local Navigation Aid. The proposed smart vision prototype is a small, cheap and easily wearable navigation aid for blind people and people with visual impairment. Its functionality addresses global navigation for guiding the user to some destiny, and local navigation for negotiating paths, sidewalks and corridors, with avoidance of static as well as moving obstacles. Local navigation applies to both indoor and outdoor situations. In this article, they focused on local navigation: the detection of path borders and obstacles in front of the user and just beyond the reach of the white cane, such that the user can be assisted in centering on the path and be alerted to looming hazards. Using a stereo camera worn at chest height, a portable computer in a shoulder strapped pouch or pocket, and only one earphone or small speaker, the system is inconspicuous, it is no hindrance while walking with the cane, and it does not block normal surround sounds. The vision algorithms are optimized so that the system can work at a few frames per second.

Brilhault, A. in 2011 introduced Pedestrian Positioning. They designed an assistive device for blind people based on adapted GIS, and fusion of GPS and vision-based positioning. The proposed assistive device may improve user positioning, even in urban environments where GPS signals are degraded. The estimated position would then be compatible with assisted navigation for blind people. Interestingly, the vision module may also answer needs of blind people by providing them with situational awareness (localizing objects of interest) along the path. Note that the solution proposed for positioning could also enhance autonomous robot or vehicle localization.

Bharat Bhargava in 2011 explored Pedestrian Crossing, a mobile cloud collaborative approach provided here for context-aware outdoor navigation, where the computational power of resources is made available by cloud computing providers for real-time image processing. The system architecture also has the advantage of being extensible and having minimal infrastructural reliance, thus allowing for wide usability. Researchers developed an outdoor navigation application with integrated support for pedestrian crossing guidance and reported experiment results, which suggest that the proposed approach is promising for real-time crossing guidance for blind people.

Boris Schauertey in 2012 proposed Find Lost Things. They proposed a computer vision system that helps blind people to find lost objects. To this end, they combined color and SIFT-based object detection with signification to guide the hand of the user toward potential target object locations. The product is able to guide the user's attention and effectively reduce the space in the environment that needs to be explored. Researchers verified the suitability of the proposed system in a user study. They experimentally demonstrated that the system makes it easier for users with visual impairments to find misplaced items, especially if the target object is located at an unexpected location.

Alexander Dreyer Johnsen in 2012 proposed touch-based mobile for users with visual impairments. They revealed the idea of using touch-based phones for users with visual impairments. Two possible technologies are screen readers and haptics (tactile feedback). In this paper, they suggested a solution based on a combination of voice and haptics. Design research was chosen as the methodology for the project. It highlights the importance of developing a solution over the course of several iterations and performing product evaluation using external participants. The research contribution is an Android-based prototype that demonstrates the new user interface, allowing the user with visual impairments to seamlessly interact with a smartphone. Operation relies on voice and haptic feedback, where the user receives information when tapping or dragging the finger across the screen. The proposed solution is unique in several ways: it keeps gestures to a minimum, it does not rely on physical keys and it takes the technologies of screen readers and haptics one step further.

K. Matusiak in June 2013 invented object recognition in mobile phones. Researchers described main features of software modules developed for Android smartphones that are dedicated to blind users. The main module can recognize and match scanned objects to a database of objects (e.g., food, medicine containers). The two other modules are capable of detecting major colors and locating directions of the maximum brightness regions in the captured scenes. The paper included a short summary of the tests of the software-aiding activities of daily living of a blind user.

12.3 Modeling of the Product

Our biggest priority when designing the product was to make it economical without losing any efficiency. The product is designed to have a user-friendly interface so that any non-technical end user can use it comfortably. The main idea of the device is to have three blocks: input device, processing device and output device. The device takes real-time video as input. Then, it processes those videos instantaneously and detects various objects and obstructions. Based on these detections, the product generates a knowledge base that is interpreted, and the output is converted to speech signals that are transmitted back to the user using a microphone. The input unit consists of two cameras mounted on spectacles. This will help users to wear the camera easily. The processing unit is a handheld box consisting of a Rasphberry Pi block that does the processing of

FIGURE 12.1
Basic outline of the device.

the video and object detection. It also does the speech conversion. The speech output is transmitted to the user via the earphones. The basic block diagram of the product is given in Figure 12.1. Apart from this, there is a power block consisting of batteries to run the system.

12.3.1 Hardware Required

The hardware requirements of the device are as follows:

1. Input Block:
   - USB Cc

2. Processing Block:
   - Raspberry Pi 3 B+1.2 GHz processor with minimum 1 GB RAM is needed to support the custom algorithm.

3. Power Block:
   - SB power bank: Any 5 V DC power source will suffice. Power consumption of the system is about 7.5 watt.

4. Output Block:
   - Bluetooth earphone/aux output (can be configured at the time of deployment)

12.3.2 Construct of Device

The device is constructed by connecting the various hardware devices as mentioned above. The figure shows the various components of the device and the connections between them.

Figure 12.2, shows there components of the device: Raspberry Pi 3 B+, earphones and power bank. The Raspberry Pi 3 B+ is the main processing unit that runs the algorithms for video processing, object detection and speech generation. The component of output unit, the earphones, is shown to be attached to the 3.5 mm mic port of the Raspberry Pi 3 processor. The speech signal generated by the algorithms will be transmitted to the user by these earphones. The other component is the power bank, which supplies 5 V power to the processor. This forms the power unit of the device.

Figure 12.3 shows the input component, a simple eye piece or spectacle with two cameras mounted—one at the horizontal level for checking the ambient environment and another vertically mounted for checking ground clearance. The user can wear the eye piece using straps at the back while navigating.

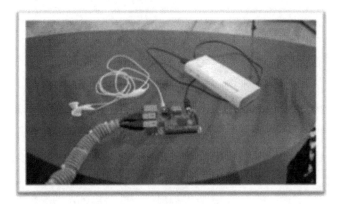

FIGURE 12.2
Components – Raspberry Pi 3 B+, Earphone, Power bank.

FIGURE 12.3
Eye piece.

Figure 12.4 shows the back side of the eye piece. A foam piece is attached at the back of the eyepiece for the user's comfort. The strap for wearing the eye piece is also attached, which will keep the eye wear in the correct position.

Figure 12.5 shows a close view of the main processor board, Raspberry Pi processor board. Various connections of the other components can be seen. The two black USB connections, on the left side, are of the two USB cameras of the input component. The white 3.5 mm connection, on the top left, is the earphone attached. This forms the output unit. On the top right, the black wire is connected to the power bank (white in color). This wire connects the power unit to the processing unit.

12.4 Algorithms

The device executes various algorithms at different stages to arrive at the required output. Various algorithms, such as ground and obstruction detection, human detection, K-means algorithm, door detection and speech generation, are executed in sequence to

FIGURE 12.4
Eye piece (back side).

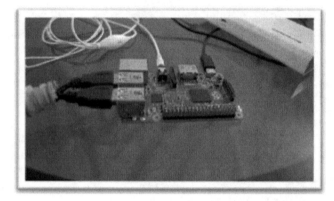

FIGURE 12.5
Raspberry Pi 3 B+ processor board.

achieve the desired result. The flowchart showing the work flow of various algorithms is given in Figure 12.6.

To start utilizing the benefits of the device, the user powers on the device and lets the OS boot up for the first time. The application starts after the boot-up process is complete. For the first time, the software needs to be calibrated according to the height of the user in order to identify possible interested regions within a captured frame (e.g. amount of ground area to be present for subsequent movement). Once calibrated, the hardware works independently and starts taking snaps of the environment continuously and processes the snaps to determine a conclusion about whether the path in front is safe or not. If not, then it finds the direction where a most probable safe path exists, depending on the clearance in front of user. The device doesn't classify all the objects in front but does classify some special cases that seem to play an important role in navigation of a person with visual impairment (e.g., doors, staircases, humans, ground). The feedback provided to the user is based on speech signals. According to previous medical research, for any person with visual impairment the ears are the next powerful sense organs, so voice feedback is quite reliable. But keeping the ears busy all the time with speech output is not a great advantage; rather, it will interfere with the natural navigation system of the

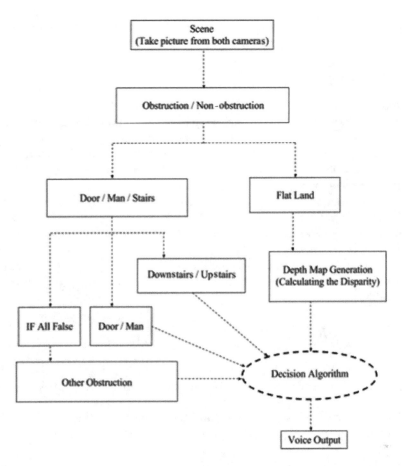

FIGURE 12.6
Process flow diagram.

human body. Whenever a major change in environment is observed, this information is given to the user in the form of speech signals. A major change in the environment includes a staircase or door, which does not allow a natural flow of walk, thus enabling the user to learn about the obstacles beforehand.

The flow of the task that is performed is as follows:

- Acquire image from the camera.
- Send the digital image to the device via USB cable or Bluetooth.
- Next is object segmentation.
- After the classes of objects are segregated out, the next task is to extract the features from them based on predefined criteria.
- The features that are extracted form the base for classification of the particular object into door, staircase, human or open ground.
- Once the classification is done, a judgment algorithm is run to decide whether a major change has been observed in the environment or not. If observed, the user is made aware of the change in the form of a suitable verbal phrase he has selected during installation of the system.

The previous steps are followed repeatedly until the device is switched off or the battery has been drained.

12.4.1 Ground and Obstacle Detection

The first step of a sequence of algorithms is to detect obstacles from one captured image. All images are processed as grayscale images in this study because the information quantity can be reduced to one third of color images, and the color information is unlikely to be crucial. Since the natural environment is difficult to recognize with only one criterion, the obstacle detection method described in this section combines two fundamental image processes: edge extraction and thresholding [4].

The edge of the image represents the outline of obstacles, and the surface and shadow of obstacles have brighter or darker pixels. The derived edge and surface extraction results are integrated in order to compare one result with the other to determine if there are any errors.

The edge of obstacles is extracted by calculating the variance in a small square window that has a width of w[pixel]. The variance value V (p, q) at (p, q) on the image is calculated as follows:

$$V(p, q) = \frac{1}{w^2} \sum_{i=p-\frac{w}{2}}^{p+\frac{w}{2}} \sum_{j=q-\frac{w}{2}}^{q+\frac{w}{2}} (A(p, q) - x(i, j))^2 \qquad 12.1$$

where A(p, q) is the average of pixel values in the window, and x(i, j) is the pixel value at (i, j).

The variance is calculated on each pixel, and then binaries by the threshold "T_h", which is one of the unique points in the proposed obstacle detection method, are given as follows:

$$T_h = \overline{V(p, q)} + t\sigma \qquad 12.2$$

where σ is a standard deviation of V (p, q), and t is an experimentally decided constant. By this expression, the threshold, which is the key parameter in the binarizing process, is decided according to the captured image.

This thresholding process expects to properly detect obstacle areas even in natural and unknown environments because the algorithm is robust to the change in light conditions. The surface of obstacles is detected by extracting bright and dark pixels. Thresholds are derived in the same way as Equation 12.2 in order to secure the robustness. Finally, these results are mixed inclusively, and areas that do not have both edge and surface information simultaneously are rejected, because such areas tend to represent no obstacles but rather tiny roughness of the surface.

12.4.2 Modified K-Means Algorithm

The primary idea of this algorithm is to detect the peaks in the histogram and to employ the corresponding gray levels as initial cluster centroids for k-means clustering. For the first centroid, the gray level of the highest peak in the histogram is selected. Then, the local maximum with the largest weighted distance to all other known centroids is chosen as a new centroid [5,6].

For the distance metric, the product of the gray level difference $d_{n,i}$ between centroid C_n and local maximum i, and the height hi of the local maximum is used. This process is iteratively repeated until the desired number of centroids is found. The product (instead of the sum) of all weighted distances is used as a criterion, since if all ($h_i.d_{n,i}$) were large, the product would be maximized [7]. The flowchart is as follows:

1. Calculate the image histogram.
2. Detect all local maxima in the histogram. We assume a total number of I local maxima are detected.
3. Choose the global maximum of the histogram as the first centroid C1.
4. Identify the remaining centroids as follows:
   a. Calculate

$$P_{N+1}(i) = \prod_{n=1}^{N} (h_i * d_{n,i})$$

   for each local maximum (i = 1, 2, ..., I). N is the number of already-found centroids. The computational complexity can be reduced by using a recursive implementation:

$$P_2(i) = h_i * d_{1,i}$$

$$^P N + 1^{(i)} = {}^P N^{(i)} * (h_i * d_{N,i})$$

   which@@@ employs a similar idea of Viterbi algorithm.
   b. Then, the new centroid $C_{N+1} = \text{argmax}(P_{N+1}(i))$ N = N + 1.
5. Repeat step 4 until N equals the user-specified number of classes K.

12.4.3 Human Detection

There are three main contributions to our object detection framework. We will introduce each of these ideas briefly below and then describe them in detail in subsequent sections. The first contribution is a new image representation called an integral image that allows for very fast feature evaluation. Motivated in part by the work of Papageorgiou et al. our detection system does not work directly with image intensities. Like these authors, we use a set of features that are reminiscent of Haar Basis functions (though we will also use related filters that are more complex than Haar filters). In order to compute these features very rapidly at many scales, we introduce the integral image representation for images. The integral image can be computed from an image using a few operations per pixel. Once computed, any one of these Haarlike features can be computed at any scale or location in constant time. The second contribution of this paper is a method for constructing a classifier by selecting a small number of important features using AdaBoost. Within any image subwindow the total number of Haarlike features is very large, far larger than the number of pixels. In order to ensure fast classification, the learning process must exclude a large majority of the available features and focus on a small set of critical features. Motivated by the work of Tieu and Viola, feature selection is achieved through a simple modification of the AdaBoost procedure: the weak learner is constrained so that each weak classifier returned can depend on only a single feature. As a result, each stage of the boosting process, which selects a new weak classifier, can be viewed as a feature selection process. AdaBoost provides an effective learning algorithm and strong bounds on generalization performance [8].

The third major contribution of this paper is a method for combining successively more complex classifiers in a cascade structure that dramatically increases the speed of the detector by focusing attention on promising regions of the image. The notion behind focus of attention approaches is that it is often possible to rapidly determine where in an image an object might occur. More complex processing is reserved only for these promising regions. The key measure of such an approach is the "false negative" rate of the attentional process. All, or almost all, object instances must be selected by the attentional filter.

We will describe a process for training an extremely simple and efficient classifier that can be used as a "supervised" focus of attention operator. The term "supervised" refers to the fact that the attentional operator is trained to detect examples of a particular class. In the domain of face detection, it is possible to achieve fewer than 1% false negatives and 40% false positives using a classifier constructed from two Harrlike features. The effect of this filter is to reduce by over one half the number of locations where the final detector must be evaluated. Those sub-windows that are not rejected by the initial classifier are processed by a sequence of classifiers, each slightly more complex than the last. If any classifier rejects the sub-window, no further processing is performed. The structure of the cascaded detection process is essentially that of a degenerate decision tree and, as such, is related to the work of Geman and colleagues.

This object detection procedure classifies images based on the value of simple features. The most common reason for using features rather than the pixels directly is that features can act to encode ad-hoc domain knowledge that is difficult to learn by using a finite quantity of training data.

The simple features used are reminiscent of Haar basis functions that have been used by Papageorgiou et al. More specifically, we use three kinds of features. The value of a two rectangle feature is the difference between the sum of the pixels within two rectangular regions. The regions have the same size and shape and are horizontally or vertically adjacent (see Figure 12.7). A three rectangle feature computes the sum within

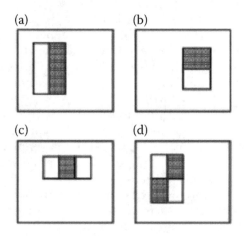

FIGURE 12.7
Example rectangle features shown relative to the enclosing detection window: (a) and (b) shows two rectangle features, (c) shows a three rectangle feature, and (d) a four rectangle feature.

two outside rectangles subtracted from the sum in a center rectangle. Finally, a four rectangle feature computes the difference between diagonal pairs of rectangles.

The sum of the pixels that lie within the white rectangles are subtracted from the sum of pixels in the grey rectangles. Two rectangle features are shown in Figure 12.7 (a) and (b); (c) shows a three rectangle feature, and (d) a four rectangle feature.

Rectangle features can be computed very rapidly using an intermediate representation for the image that we call the integral image. The integral image at location (x,y) contains the sum of the pixels above and to the left of (x,y) inclusive:

$$ii(x, y) = \sum_{x' \leq x, y' \leq y} i(x', y')$$

The value of the integral image at location 1 is the sum of the pixels in rectangle A. The value of location 2 is $A + B$, at location 3 is $A + C$, and at location 4 is $A + B + C + D$. The sum within D can be computed as $4 + 1(2 + 3)$, where ii(x,y) is the integral image and i(x,y) is the original image. Using the following pair of recurrences:

$$s(x, y) = s(x, y - 1) + i(x, y)$$
$$ii(x, y) = ii(x - 1, y) + s(x, y)$$

where s(x,y) is the cumulative row sum, $s(x, -1) = 0$ and $ii(-1, y) = 0$, the integral image can be computed in one pass over the original image.

Using the integral image, any rectangular sum can be computed in four array references (see Figure 12.8). Clearly, the difference between two rectangular sums can be computed in eight references. Since the two-rectangle features defined above involve adjacent rectangular sums they can be computed in six array references, eight in the case of the three rectangle features, and nine for four rectangle features.

Rectangle features are somewhat primitive when compared with alternatives such as steerable filters. Steerable filters, and their relatives, are excellent for the detailed analysis of boundaries, image compression and texture analysis. In contrast, rectangle features—while sensitive to the presence of edges, bars and another simple image structure—are quite coarse. Unlike steerable filters, the only orientations available are vertical, horizontal and diagonal. The set of rectangle features do however provide a rich image representation that supports effective learning. In conjunction with the integral image, the efficiency of the rectangle feature set provides ample compensation for their limited flexibility.

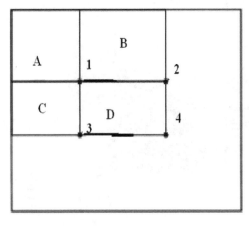

FIGURE 12.8
The sum of the pixels within a rectangle computed with four array references.

Given a feature set and a training set of positive and negative images, any number of machine learning approaches could be used to learn a classification function. In our system, a variant of AdaBoost is used both to select a small set of features and to train the classifier. In its original form, the AdaBoost learning algorithm is used to boost the classification performance of a simple (sometimes called weak) learning algorithm. There are a number of formal guarantees provided by the AdaBoost learning procedure. Freund and Schapire proved that the training error of the strong classifier approaches zero exponentially in the number of rounds. More importantly, a number of results were later proved about generalization performance. The key insight is that generalization performance is related to the margin of the examples and that AdaBoost achieves large margins rapidly. The AdaBoost algorithm for classifier learning is as follows (Figure 12.9):

- Given example images $(x_1, y_1), ..., (x_n, y_n)$ where $y_i = 0, 1$ for negative and positive examples respectively.
- Initialize weights $w_{1,i} = \frac{1}{2m}, \frac{1}{2l}$ for $y_i = 0, 1$ respectively, where m and l are the number of negatives and positives respectively.
- For $t = 1, ..., T$
  1. Normalize the weights, $w_{t,i} \leftarrow \frac{w_{t,i}}{\sum_{j=1}^{n} w_{t,j}}$, so that w_t is a probability distribution.
  2. For each feature, j, train a classifier h_j that is restricted to using a single feature. The error is evaluated with respect to w_t, $\varepsilon_j = \sum_i w_i |h_j(x_i) - y_i|$
  3. Choose the classifier, h_t, with the lowest error ε_t.
  4. Update the weights: $w_{t+1,i} = w_{t,i} \beta_t^{1-\varepsilon_i}$

where $\varepsilon_i = 0$ if example x_i is classified correctly, $\varepsilon_i = 1$ otherwise, and $\beta_t = \frac{\varepsilon t}{1 - \varepsilon t}$.
The final strong classifier is:

$$h(x) = \begin{array}{l} 1 \quad \sum_{t=1}^{T} \alpha_t h_t(x) \geq \sum_{t=1}^{T} \alpha_t \\ 0 \quad otherwise \end{array}$$

The two features are shown in the top row and then overlaid on a typical training face in the bottom row. The first feature measures the difference in intensity between the region

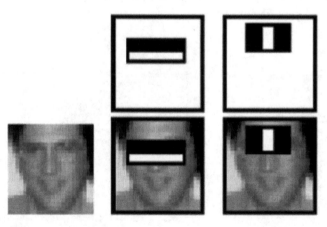

FIGURE 12.9
The first and second features selected by AdaBoost.

of the eyes and a region across the upper cheeks. The feature capitalizes on the observation that the eye region is often darker than the cheeks. The second feature compares the intensities in the eye regions to the intensity across the bridge of the nose.

The overall form of the detection process is that of a degenerate decision tree, what we call a "cascade". A positive result from the first classifier triggers the evaluation of a second classifier that has also been adjusted to achieve very high detection rates. A positive result from the second classifier triggers a third classifier, and so on. A negative outcome at any point leads to the immediate rejection of the sub window.

Stages in the cascade are constructed by training classifiers using AdaBoost and then adjusting the threshold to minimize false negatives. Note that the default AdaBoost threshold is designed to yield a low error rate on the training data. In general, a lower threshold yields higher detection rates and higher false positive rates.

Computation of the two-feature classifier amounts to about 60 microprocessor instructions. It seems hard to imagine that any simpler filter could achieve higher rejection rates. By comparison, scanning a simple image template, or a single layer perceptron, would require at least 20 times as many operations per sub-window.

The structure of the cascade reflects the fact that within any single image an overwhelming majority of sub windows are negative. As such, the cascade attempts to reject as many negatives as possible at the earliest stage possible. Although a positive instance will trigger the evaluation of every classifier in the cascade, this is an exceedingly rare event.

Much like a decision tree, subsequent classifiers are trained using examples that pass through all the previous stages. As a result, the second classifier faces a more difficult task than the first. The examples that make it through the first stage are "harder" than typical examples. The more difficult examples faced by deeper classifiers push the entire receiver operating characteristic (ROC) curve downward. At a given detection rate, deeper classifiers have correspondingly higher false positive rates.

The cascade training process involves two types of trade-offs. In most cases, classifiers with more features will achieve higher detection rates and lower false positive rates. At the same time, classifiers with more features require more time to compute. In principle, one could define an optimization framework in which:

a. the number of classifier stages,

b. the number of features in each stage, and

c. the threshold of each stage,

are traded off in order to minimize the expected number of evaluated features. Unfortunately, finding this optimum is a tremendously difficult problem.

In practice, a very simple framework is used to produce an effective classifier that is highly efficient. Each stage in the cascade reduces the false positive rate and decreases the detection rate. A target is selected for the minimum reduction in false positives and the maximum decrease in detection. Each stage is trained by adding features until the target detection and false positives rates are met. (These rates are determined by testing the detector on a validation set.) Stages are added until the overall target for false positive and detection rate is met.

The complete face detection cascade has 38 stages with over 6,000 features. Nevertheless, the cascade structure results in fast average detection times. On a difficult dataset, containing 507 faces and 75 million sub-windows, faces are detected using an average of 10 feature evaluations per sub-window. In comparison, this system is about 15 times faster than an implementation of the detection system.

12.4.4 Door Detection

The door detection method is based on the Canny edge detection algorithm and the Hough transform, and it uses fuzzy logic to determine the probability of existence of doors that fit a set of predefined rules. The implementation of the algorithm is composed of a few core steps: region of interest (ROI) preprocessing, door line extraction and door classification.

12.4.4.1 ROI Preprocessor

The preprocessing steps locate ROIs that are likely to contain doors. Door corners are likely among the strongest corners in the image; however, regular corner detection not only detects the corners of the door, but also any other strong corners in the image. To differentiate the cases, the preprocessing algorithm takes into account that the corners for doors are near the intersection of strong horizontal and vertical lines. This sacrifices some degree of rotational invariance, but significantly reduces the number of undesirable corners [9].

The image is converted to grayscale and cropped slightly to remove border effects. Canny edge detection is then performed, and the resulting image is dilated with a 3×3 cross-structuring element to improve connections at the corners. Dilation with a horizontal and vertical line structuring element is performed before corners are computed. Region counting is then performed, and each line segment is assigned a weighting based on the region size.

Next, the corners are determined by multiplying the squares of the horizontal and vertical segment images similar to what would be performed with a Harris corner detector with a sensitivity parameter equal to zero. The corners of the door are discernibly the strongest, especially near the top where the horizontal segment is clearer in the image.

After corner detection is performed, multiple sets of three corners with an implicit fourth corner are used to generate quadrilateral regions where a door would likely be. During the initial search, the interior angles of the polygon must be within a tolerance comparable to those of a door viewed with a slightly rotated camera and have a height-to-width ratio that is acceptable. Next, the quadrilateral regions are each evaluated based on overall area and independently on a shape metric. Particular emphasis is given to quadrilaterals that could be described as rectangles and/or parallelograms and those that match the appropriate 2.2:1 standard for U.S. doors.

The quadrilaterals selected based on area are merged into one or more quadrilaterals of maximum extent. The quadrilaterals selected based on shape are evaluated for overall connection in the edge image. Each remaining quadrilateral region is converted to a rectangular region, and each region is cropped from the original image, padded and sent to the line and feature extraction algorithm. If the selected region is small, the image is up-sampled before it is sent.

The ROI preprocessor increases the robustness of the door detection algorithm and, depending on the image and parameters, can speed up detection. The detection rate increase is notable, especially for images with small doors.

12.4.4.2 Door Line and Feature Extraction

With the ROI preprocessor removing most of the context around the door, the remaining image is grayscale and processed to extract dominant line features that are put into a line database. The database extraction is a multistep process. First, the input image is converted into an edge map using Canny edge detection, with a very low fixed threshold.

Morphological erosion with 3×1 pixel line structuring elements is used to extract horizontal and vertical lines from the edges.

Region counting is applied to remove small vertical and horizontal lines from the respective images. Only regions of sufficient size are retained. A Hough transform is applied to each horizontal and vertical line. The theta and rho values are stored, along with the line segment locations, in a database. The final line database consists of horizontal and vertical lines that meet a length criterion and an angle criterion.

Eight-level Otsu thresholding is applied to the original image passed by the ROI preprocessor. Note that the Otsu thresholding function was leveraged. Each Otsu level is turned into a binary image with all levels below the current level set to 0 and all levels equal to or larger than the current level set to 1. Each binary image is region-counted looking for features that are the correct size for hinges or door knobs.

The hinge locations are stored in a database, whereas the door knobs are stored in a separate database. The result is then passed onto an algorithm utilizing fuzzy logic to calculate confidence.

12.4.4.3 Fuzzy Logic to Extract Doors

A heuristic method is used to extract parallel lines from the line database. The parallel lines are used to find the four corners of the door.

The heuristic method employs three metrics to determine the likelihood of line segments being part of a door. One metric is assigned to vertical line pairs, the second metric is assigned to the top horizontal line, and the third metric is assigned to the bottom horizontal line. High metric values indicate more confidence that the line combination constitutes a door. The steps used to calculate vertical line metrics are as follows:

Calculate the mean length of each pair of vertical line segments. If the distance between the pair of lines is greater than one third and less than two thirds of the mean length, the pair is considered further. The difference between the two vertical lines is subtracted from the metric. If the distance between the vertical lines multiplied by 2.2 is close to the length of the vertical line pair, the metric is increased. Vertical line pairs with a metric less than 50% of the highest vertical line metric are removed from consideration.

Once the vertical line pairs are scored, the fuzzy logic scores the horizontal lines with respect to each pair of vertical lines still under consideration. The average top and bottom Y coordinates of vertical lines are calculated. Each horizontal line has a metric calculated based on how close the average Y coordinate of the horizontal line matches the top or bottom average Y values of the vertical line pair. The lines with the highest metric are stored as the top and bottom that match with the vertical line pair. Horizontal lines that do not span the entire distance between the two vertical lines score a lower horizontal metric.

Once all three base metrics are calculated, the four lines constituting the door are passed to the door feature detection section of the fuzzy logic. The door feature detection section uses the location of the door as described by the four lines to search for features that improve the confidence score. The door feature code works as follows. A door handle near either vertical line in the second quadrant from the bottom of the door adds half the difference between the vertical line score and the top and bottom horizontal line scores independently. Hinges near either vertical line add a sixth to the difference between the vertical line score and top and bottom horizontal line scores independently. The door feature detection helps significantly with close door detection, which is defined as doors closer than 6 ft. This is because the top or bottom of the door is commonly missing from these images. However, door handles and hinges become more easily identifiable with thresholding as these features are larger in close door pictures than other picture types.

The feature metric values are summed with the original vertical and horizontal door metrics to create the final door confidence metric. These final values are sorted from highest to lowest. The last section of door confidence is determining if the metrics constitute a door, and how many. The highest ranked combined metric is indicative of the existence of a door, if its horizontal metrics are at least one fourth of the vertical line metric. In addition, all lower confidence doors are thrown away if their area overlaps 50% or more of the higher scored rectangle. If the top few candidates have close metrics and their areas do not overlap, it is indicative of multiple doors in the image.

The overall algorithm of sensing a door is as follows:

1. Obtain the high pass filtered image such that the edges are obtained.
2. Count the number of white-colored pixels and black-colored pixels.
3. Now, the object of interest is generated based on the minimum variance obtained.
4. Apply vertical Hough transform to the image.
5. Set the visited array as 0.
6. Move along the image toward the right and check if the transition occurs from black pixel to white pixel. If it occurs, mark the first visited node. Again, check for the transition from white pixel to black pixel. If it occurs, mark the second visited node.
7. If both visited nodes are marked, then sum up the difference between the visited node x coordinates.
8. Repeat steps 5 and 6 until the condition in step 6 is satisfied or the image frame ends.
9. Generate the mean.
10. Repeat steps 4, 5, and 6 to obtain the minimum square error for the object.
11. (Mean square error/mean) The white pixel count forms the feature set for door detection.

$$\text{Mean}(M) = \frac{\Sigma_{i=1,B}\,(X_{2i} - X_{1i})}{|Y_K - Y_j|}$$

$$\text{Mean Square Error}(MSE) = \frac{\Sigma(M - (X_{2i} - X_{1i}))^2}{|Y_K - Y_j|}$$

12.4.5 Stair Detection

The algorithm for sensing the presence of stairs is as follows:

1. Obtain the high pass filter image.
   A high pass filter is the basis for most sharpening methods. An image is sharpened when contrast is enhanced between adjoining areas with little variation in brightness or darkness.
   These filters emphasize fine details in the image—exactly the opposite of the low pass filter. High pass filtering works in exactly the same way as low pass filtering; it just uses a different convolution kernel.
2. Apply horizontal Hough transform to the image.
3. Scan the image through the middle of the frame with a predefined kernel of dimension 4x3.
4. Generate two variables, $Temp_{var1}$ & $Temp_{var2}$. The first one gives the correlation with the kernel middle values. The second one gives the correlation with the entire kernel at each step.
5. If $Temp_{var1} = 6$ and $5 < Temp_{var2} < 7$, then increment $Stair_{count}$.
6. If $Stair_{count} > 1$, then scan again through the middle of the frame and obtain the coordinates of the point of interest for the kernel.
7. Get the difference between consecutive coordinate values and store them as the step height.
8. Get the mean step height by using $\frac{Step\ Height}{Step\ Count}$ and mode step height ($Step_{Height}$ mostly occurring).
9. If the mean step height deviates largely from the mode step height, then it is not a stair. Otherwise, it is a stair.

Kernel Used:

0	0	0	0
1	1	1	1
1	1	1	1
0	0	0	0

12.5 Output

The outputs are shown for various cases such as door, obstructions, human, stairs, etc.

12.5.1 Door Detection

The door detection algorithm separates the door from the background and finds the optimum distance from the person wearing the device. A secondary camera is there to find the clear portion (green patch) of floor for effective locomotion (Figure 12.10).

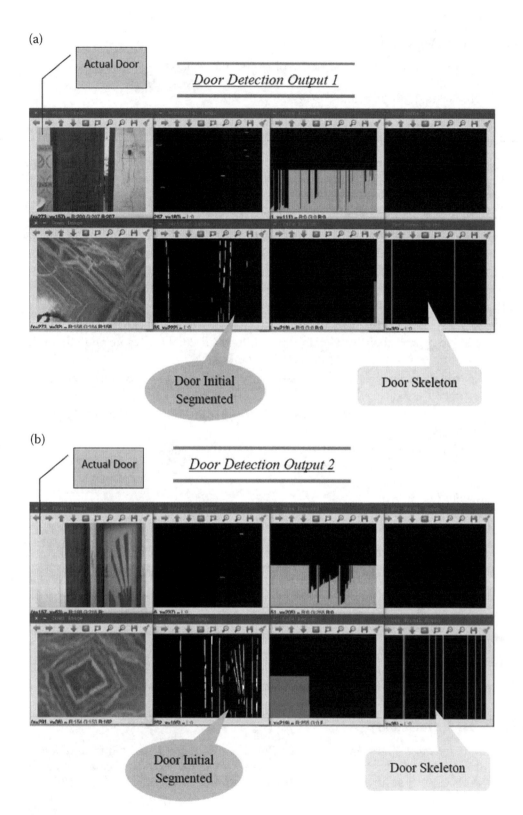

FIGURE 12.10
(a) Door Detection 1 (b) Door Detection 2.

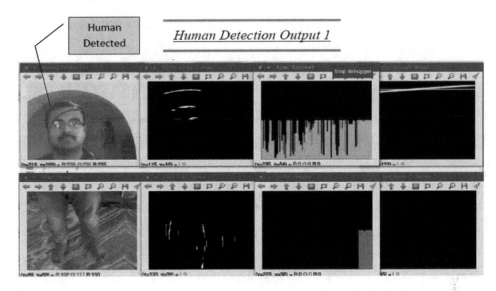

FIGURE 12.11
Human Detection 1.

12.5.2 Human Detection

The human detection algorithm finds the face of the person interacting and provides an audio feedback of the person's history (unknown or already registered) (Figure 12.11, Figure 12.12).

12.5.3 Stair Detection

The staircase detection algorithm finds the sharp edges of the stairs and uses projections into the horizontal plane to find the optimum distance and response time to reach that stair (Figure 12.13, Figure 12.14).

12.6 Conclusion

This product is an amalgamation of video processing, object detection and speech generation. This device is a build-up of various devices connected together to function in unison and generate the desired output. Cameras take real-time video. Raspberry Pi processor processes the video to detect humans, objects, stairs and doors. Finally, the findings are converted to speech and transmitted by microphone to the user. A number of state-of-the-art algorithms work to detect various objects precisely. Multiple trial runs have shown that the device is successful in more than 98% of cases in indoor setup consisting of humans, doors, stairs and obstructions. Even in outdoor setup, trials shows a success rate of approximately 95% of cases. The main risk in using the device in an outdoor setting is the time lapse in detecting moving objects such as cars. This is the challenge that needs to be overcome. However, in the current scenario, this device can greatly help persons with visual impairments move indoors. They can move

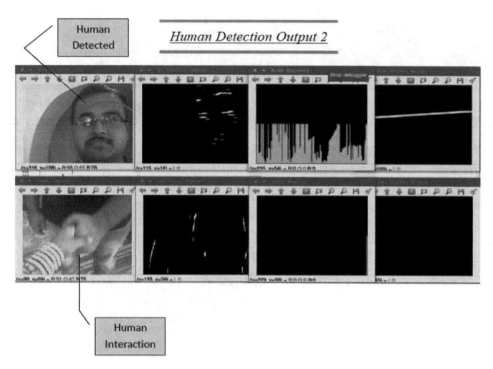

FIGURE 12.12
Human Detection 2.

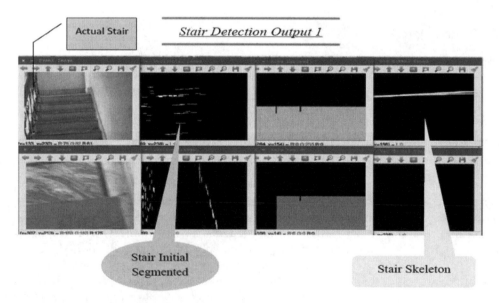

FIGURE 12.13
Stair Detection 1.

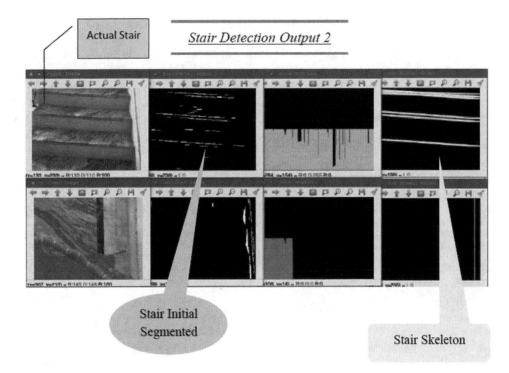

FIGURE 12.14
Stair Detection 2.

independently and confidently with the use of this device. The device is portable and low cost. Future research can be directed to real-time detection of fast-moving objects so that this device can be used in outdoor conditions as well. Moreover, the movement of the person can be uploaded in real time via a GPS system so that his or her movements can be monitored remotely and his or her safety can be ensured.

References

1. G. Heitz, S. Gould, A. Saxena, D. Koller (2008), Cascaded classification models: Combining models for holistic scene understanding. *Advances in Neural Information Processing Systems 21 (NIPS 2008)*.
2. G. Bradski, A. Kaehler (2008), *Learning OpenCV: Computer Vision with the OpenCV Library*, O'Reilly, USA.
3. T. L. Campbell, E. K. Karichet (2014), Mobility device and method for guiding the visually impaired, United States patent, Patent no.: US 8,825,389 B1, USA.
4. G. M. Elchinger (1981), Mobility cane for the blind incorporating ultrasonic obstacle sensing apparatus. Patent no. US 4280204 A, USA.
5. S. Ray, R. H. Turi (2000), Determination of number of clusters in K-means clustering and application in colour image segmentation (Avaliable in http://users.monash.edu).
6. T., Kanungo, C. A., San Jose, D. M., Mount N. S., Netanyahu C. D., Piatko R., Silverman A. Y., Wu (2002), An efficient k means clustering algorithm: analysis and implementation. Pattern

analysis and machine intelligence, *IEEE Transactions*, vol. 24, no. 7, pp. 881–892. 10.1109/ TPAMI.2002.1017616

7. T. Cover, P. Hart (1967), "Nearest neighbor pattern classification," *IEEE Transactions on Information Theory*, vol. 13, no. 1, pp. 21–27, doi: 10.1109/TIT.1967.1053964.

8. P. Cheeseman, M. Self, J. Kelly, J. Stutz. (1988), Bayesian classification Seventh National Conference on Artificial Intelligence, In Proceedings of the Fifth International Conference on Machine Learning, 1988.

9. T. Acharya, A. K. Ray (2005), *Image Processing: Principles and Applications*, Wiley Publication.

13

Analysis of Human Emotion-Based Data Using MIoT Technique

Stobak Dutta[1], Sachikanta Dash[2] and Neelamadhab Padhy[3]
[1]*UEM, Kolkata & GIET University (PhD Scholar), Gunupur, Odisha, India*
[2]*DRIEMS Autonomous Engineering College, Cuttack, India*
[3]*GIET University, Gunupur, Odisha, India*

CONTENTS

13.1 Introduction

A large number of people participating in a social network can give more specific answers to complex problems than a single individual or a small group of even knowledgeable persons [1]. Many researchers are investigating the potential of integrating social networking concepts with the traditional Internet of Things (IoT) concept to give rise to a Social Internet of Things (SIoT). In the SIoT paradigm, humans can impose rules on objects, and autonomous inter-object interactions can occur as a social network [2]. In this context, it is not the smart object but rather the social object that is smart that will make some difference in the advancement of healthcare [3]. The primary goal of this chapter is to explain how to monitor patients and predict their health status. One approach in this work is to discover the activity of patients using sensors. However, this process requires many sensors to document the patient's condition, which can be expensive and tiresome. The goal of this method is to detect the current status of the patient with the help of some sensors in the SIoT.

DOI: 10.1201/9780429318078-13

13.2 IoT and the SIoT Concept

The SIoT is an IoT where things can build social relations and are self-sufficient in regards to humans [4,5]. Along these lines, there exists a casual and formal method for interfacing the hubs to send and get information inside a system and association [6]. In SIoT, the "things" are social only with one another. So, we can, without much of a stretch, change a "thing" into a "smart and social" thing, building up a community and commutative system by utilizing the idea of "Social Internet of Things" [7]. For portraying the proposed framework, we will take the assistance of the three-layer design model for IoT introduced by Zheng et al. (2011) [8]. The model comprises (i) the sensing layer, which is committed to the information acquisition and hub relationship in short range and neighborhood systems; (ii) the network layer, whose capacity is to move information to various layers; and (iii) the application layer, where the various applications have a relationship with the middle ware. The three fundamental components of the proposed framework are: the SIoT server, the gateway and the object. The proposed design follows the three-layer model: the detecting, system and application layers. The segments of the application layer are the relationship management (RM), service discovery (SD), service composition (SC), and trustworthiness management (TM) functionalities. In both the object and the gateway there are some optional layers.

From the analysis of possible service and application there can be many relationship that can be built upon after analyzing the way or the relationship by virtue of how social things are connected. The kinds of relationships we define are summarized here [9,10]: "Parental object relationship" (POR): this is the relationship among the objects depending on their make. "Co-location object relationship" (C-LOR): these are the homogeneous or heterogeneous objects that exist in the same place. This relationship is mainly used in the smart city, and it is not mandatory that all devices work with the same set of protocols or for a single goal. "Co-work object relationship" (C-WOR): this relationship is mainly established whenever the objects act as a team to provide a common IoT application. In such cases, the message passing takes place between the objects, and the result is the social smart object finalizes the resultant output. "Ownership object relationship" (OOR): established among heterogeneous objects that belong to the same user.

13.3 SIoT in Healthcare

Many IoT devices are wearable and can be made social to form an SIoT network, which mainly helps to monitor and control the health metrics of patients. An example of the benefits is that a patient as well as the physician will know the current condition of the patient at any time, which will help the medical professional to take prompt action. Such devices can have a huge role in saving the life of a patient during an emergency. Connected health is emerging as a major application for developing technologies. Both commercial companies and individuals can benefit from the innovation of connected healthcare systems and smart embedded SIoT devices. The main objective is to use the research performed on new technologies to allow the formation, improvement and development of connected health systems with the idea of developing a system that can help patients obtain better awareness of their health status and receive early medical warnings.

13.4 What Is Galvanic Skin Response?

The most observable reason why sweat glands change their activity is thermoregulation of the body. When a person is performing hard work or experiencing a hot day, the body will profusely sweat to maintain body temperature, but there is another reason why the sweat glands change their activity. The perspiration organs are constrained by the sympathetic nervous system (SNS), a subdivision of the autonomic nervous system (ANS), liable for the "fight-or-flight response". Each time the body gets a tactile upgrade that could impact its resting state, the SNS enacts a physiological reaction that incorporates an incremental sweat organ action. This autonomic reaction is activated by an enthusiastic response, for example, shock, fear or outrage—or when a person is under pressure.

The galvanic skin response (GSR) is changes in sweat gland activity in response to any type of emotion based on the environment. Consider the feelings generated when something is scary or very threatening versus feelings generated by good times. When the environment changes our mood or feelings, the sweat glands, or more specifically the eccrine sweat gland, increase their activity.

It is found in virtually all skin, with the highest density in the palms and soles, then on the head.

GSR can be positive or negative:

Positive emotions—delight, happiness, joy, enjoyment

Negative emotions—unhappiness, sorrow, fear, sadness

Different types of emotions can cause higher excitement and raise skin conduction. Our objective in our work with GSR is not to distinguish the emotions but rather to give a representation of the emotions based on their intensity. In this competitive world, stress has become a common problem. Many people are affected by stress, which can result in dementia and insomnia. When people are under stress, they may develop depression, which can be dangerous for them. Stress symptoms are often measured by clinical staff who work closely with the person to answer a checklist. This process requires a lot of effort as well as continuous observation of the patient for a long time.

We introduce a stress sensor based on GSR, also known as skin conductance, which will be controlled by a low-cost, low-power, wireless communication network. The sensor has two electrodes that are placed on the fingers, and it sends the data via network to a coordinator that forwards it to a computer. The server similarly receives such data from the different sources (similar devices connected in the SIoT network).

After analyzing the data, we can determine what makes a person nervous. This type of experiment can be done on several sets of people. Then, this dataset can be used as a resource for predicting mental conditions under stress and favorable conditions. The dataset can also help users to monitor the situation in a more practical way and can be used by psychologists to help patients control stress.

We propose an arrangement that is easy to implement. The proposed model allows the devices of different operators or the same operator to cooperate on the basis of predefined policies. This will allow the implementation of the SIoT concept. A system structure is defined for object attributes to infer the social relationship between any two

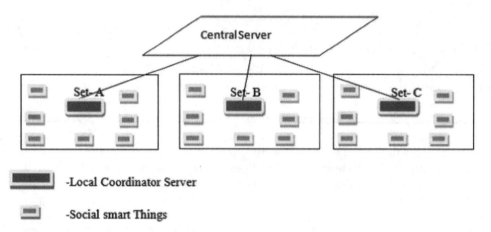

-Local Coordinator Server

-Social smart Things

FIGURE 13.1
The proposed system server node interaction.

devices in the network. In this structure, there exists a neutral server or node, which can be connected with all the nodes irrespective of their make. The server broadcasts its address periodically to all the devices in the set. If any new device is added to that set, then it will be updated with the periodic broadcast. Each node of the application set reports to the server on a definite time interval. Each node is also assigned a unique ID. The application software present in the server always keeps a back-up of all the data and sends the data to a central server. The central server connects the different servers of the different sets as shown in Figure 13.1. As shown, all the wearable SIoT devices are divided into several sets. Each set will report to its local coordinator server. The central server will receive data from all the coordinators, and on the basis of that data, users can perform analysis and make appropriate decisions.

13.5 Conclusion

There are many IoT applications that can be employed to serve many areas of interest in healthcare. Although these applications or things may belong to different operators, they may be employed in the same geographical area as SIoT. The model we presented here gathers data about stress from different sets of people. The central server collects the graphical data and converts it to a dataset from which users can draw conclusions about the causes of stress for different sets.

References

1. J. Surowiecki (2004), *The Wisdom of Crowds*, Doubleday.
2. Y. Saleem, N. Crespi, M. H. Rehmani, R. Copeland, D. Hussein, E. Bertin (2016), Exploitation of social IoT for recommendation services, *2016 IEEE 3rd World Forum on Internet of Things (WF-IoT)*, doi: 10.1109/WF-IoT.2016.7845500, IEEE, USA.

3. S. Dutta, S. Dash, A. Mitra (2020), A Model of Socially Connected Things for Emotion Detection, *IEEE – ICCSEA 2020*, GIET, Odisha, India.

4. V. Beltran, A. M. Ortiz, D. Hussein, N. Crespi (2014), A semantic service creation platform for Social IoT, *2014 IEEE World Forum on Internet of Things (WF-IoT)*, DOI: 10.1109/WF-IoT.2014 .6803173. Seoul, South Korea.

5. M. V. Villarejo, B. G. Zapirain, A. M. Zorrilla (2012), A stress sensor based on Galvanic Skin Response (GSR) controlled by ZigBee. *Sensors (Basel)*. vol. 12, no. 5, pp. 6075–6101. doi:10.3390/s120506075.

6. C. Marche et al. (2020), How to exploit the Social Internet of Things: Query Generation Model and Device Profiles' Dataset. *Computer Networks*, vol. 174, no. 107248. doi: 10.1016/ j.comnet.2020.107248.

7. S. Rho, Y. Chen (2018), Social Internet of Things: Applications, architectures and protocols. *Future Generation Computer Systems*, vol. 82, pp. 667–668. ISSN 0167-739X. doi: 10.1016/ j.future.2018.01.035.

8. L. Zheng, H. Zhang, W. Han, X. Zhou, J. He, Z. Zhang, Y. Gu, J. Wang (2011), Technologies applications and governance in the Internet of things, in O. Vermesan, P. Friess (eds), *Internet of Things – Global Technological and Societal Trends*, River Publishers, pp. 141–175.

9. L. Atzori, A. Iera, G. Morabito, M. Nitti (2012), The Social Internet of Things (SIoT) – When social networks meet the Internet of Things: Concept, architecture and network characterization, *Computer Networks*, vol. 56, no. 2012, pp. 3594–3608

10. Seeed Studio Bazaar, The IoT Hardware enabler. (2021). Retrieved 3 March 2021, from http://www.seeedstudio.com/depot/Grove

14

Understanding Explainable AI: Role in IoT-Based Disease Prediction and Diagnosis

Chittabrata Mal

Amity Institute of Biotechnology, Amity University Kolkata, Kolkata, India

CONTENTS

14.1 Introduction

For the last few years, artificial intelligence (AI) and machine learning (ML) have become very hot buzzwords. Recently, another word has been added to the field of computer science: Internet of Things (IoT). All these computational tools are being applied in different fields, starting from data management to disease diagnosis and healthcare

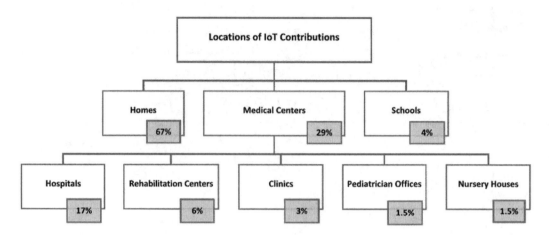

FIGURE 14.1
Proportions of the locations of IoT contributions employed in 67 selected studies. (Sadoughi et al. 2020).

management. A large number of the giant industries have already adopted such IoT-based technologies for their marketing, branding and production (Figure 14.1). Most importantly, the healthcare system has been revolutionized by the advent of IoT.

In healthcare, AI applications mainly bridge disease prediction, diagnosis, treatment approaches and patient outcomes. They can save time and costs associated with the whole process. They also help in the comprehensive analysis of huge datasets for fast and accurate decision making [1]. On the other hand, IoT coordinates with the physical objects/devices to communicate and share information with each other [2]. Thus, the traditional objects become smart by communication protocols, embedded devices, internet protocols, sensor networks and applications.

Day by day, we are becoming dependant on smart gadgets for a better life. Even un-knowingly, we are using several AI-based devices for different purposes. For example, wireless body area network (WBAN) is an emerging technology that deals with a wide range of IoT-based applications in healthcare [3]. Due to the advancement of ML and AI, WBAN attracts both academia and industry. However, certain aspects such as power resources, communication capabilities and computation power inhabit WBAN [4], which are the main challenges and open issues with ML models in such sensitive networks.

Until 2018, according to IoT Analytics, 7 billion out of over 17 billion connected devices in the world are Internet of Things (IoT) devices. The IoT is the collection of various sensors, devices and other technologies that do not directly interact with consumers the way phones or computers do. Rather, IoT devices provide information, control and analytics to connect a world of hardware devices to each other and the greater internet (Figure 14.2). With the advent of cheap sensors and low-cost connectivity, IoT devices are proliferating [5].

14.2 Limitations of Existing AI and IoT-Based Systems

Data generated by IoT-based smart devices are re-used to generate new models and to get more insight and intelligence. However, extraction of valuable information from the

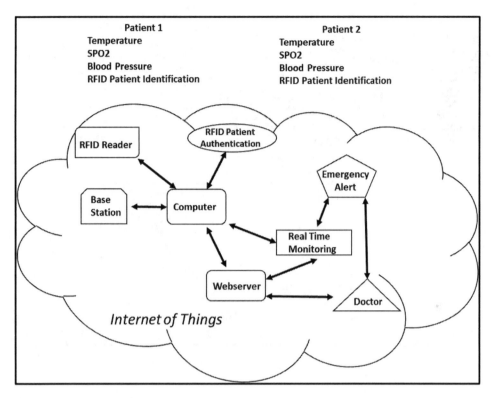

FIGURE 14.2

Physical components of the proposed healthcare information system based on an IoT architecture. RFID is an automatic identification system that includes a reader (also called interrogator) and tags. For more details, please refer the article of Aktas et al., 2018.

system and management of large data are challenging tasks. There are many aspects and subcomponents to IoT such as connectivity, security, data storage, system integration, device hardware, application development, and even networks and processes that are ever-changing in this space. Moreover, scaling of IoT functionality is also complicated. Though designing sensors that can be accessed from a smart phone is easy, creation of devices that are reliable, remotely controlled and upgraded, secure, and cost effective is much more difficult.

As healthcare providers rely more on AI, a proper trust relationship, also referred to as calibrated trust [6], becomes a requirement for effective decisions. Figure 14.3 presents an overview of important factors influencing trust in AI for healthcare, possible ways to improve trust relationships, and their impact on trust [7].

14.3 Need for Transformation of IoT by XAI

AI and IoT are combining to provide greater visibility and control of the wide array of devices and sensors connected to the internet. Most of the smart IoT devices powered by AI cannot explain why they decide a patient has a certain disease. Moreover, the physicians

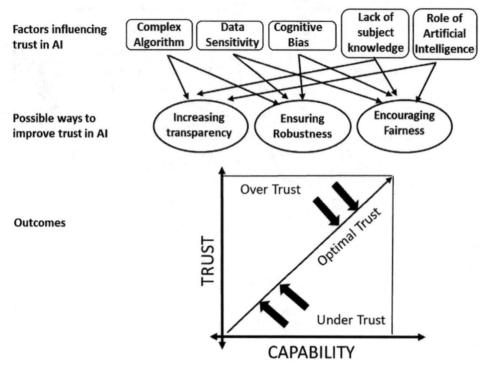

Factors influencing trust in AI

Complex Algorithm | Data Sensitivity | Cognitive Bias | Lack of subject knowledge | Role of Artificial Intelligence

Possible ways to improve trust in AI

Increasing transparency | Ensuring Robustness | Encouraging Fairness

Outcomes

Over Trust

Optimal Trust

Under Trust

TRUST

CAPABILITY

FIGURE 14.3
Human factors and trust in artificial intelligence. (Asan et al. [7]).

can not verify the predicted results easily. Thus, the acceptance of these algorithms in disease diagnostics has not increased rapidly. On the contrary, XAI is easy to understand. As a result, all signs point to XAI being rapidly adopted across healthcare, making it likely that providers will actually use the associated diagnostics [5].

14.4 What Is Explainable Artificial Intelligence (XAI) and How Does It Work?

Explainable AI (XAI) is an emerging field in machine learning that aims to address how black box decisions of AI systems are made [5,8]. This area inspects and tries to understand the steps and models involved in making decisions. XAI is thus expected to answer the following: Why did the AI system make a specific prediction or decision? Why didn't the AI system do something else? When did the AI system succeed, and when did it fail? When do AI systems give enough confidence in the decision that one can trust it, and how can the AI system correct errors that arise?

XAI can use two types of ML algorithms. Comparatively simple ML algorithms such as decision trees, Bayesian classifiers and other algorithms can be deployed in XAI systems. These are inherently explainable and also have certain amounts of traceability and transparency in their decision making without sacrificing too much

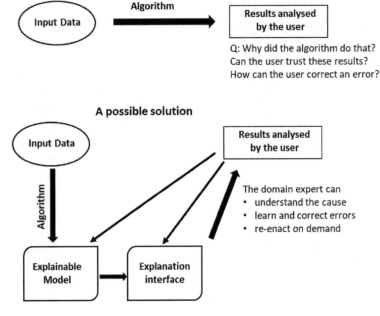

FIGURE 14.4
Concept of XAI (Holzinger et al. [10]).

performance or accuracy [5]. More complicated, but also potentially more powerful, algorithms such as neural networks, ensemble methods including random forests, and other similar algorithms sacrifice transparency and explainability for power, performance and accuracy.

Explainable models also have advanced human–machine collaboration. They allow the AI to perform tasks that it is better suited for, such as collecting and structuring information or extracting patterns from massive amounts of data. Meanwhile, humans can focus on extracting meaning from the data collected [9]. XAI models should include completeness, correctness, compactness and actionability so that medical professionals reviewing the algorithmic output and the accompanying explanation have a clear – and hopefully improved – path to clinical decisions. Moreover, we need to understand the XAI model along with an explanation interface that makes the results gained in the explainable model not only usable but also useful to the expert (Figure 14.4) [10].

14.5 Overview of XAI Models

There are two types of XAI models: (i) post-hoc explainability = "(lat.) after this", occurring after the event in question (e.g., explaining what the model predicts in terms of what is readily interpretable); (ii) ante-hoc explainability = "(lat.) before this", occurring before the event in question (e.g., incorporating explainability directly into the structure of an AI model, explainability by design).

14.5.1 Post-Hoc Systems

Post-hoc systems aim to provide local explanations for a specific decision and make it reproducible on demand (instead of explaining the whole system's behavior). A representative example is LIME (Local Interpretable Model-Agnostic Explanations), developed by Ribeiro et al. [11], which is a model-agnostic system, where $x \in R\ d$ is the original representation of an instance being explained, and $x\ 0 \in R\ d\ 0$ is used to denote a vector for its interpretable representation (e.g., x may be a feature vector containing word embeddings, with $x\ 0$ being the bag of words). The goal is to identify an interpretable model over the interpretable representation that is locally faithful to the classifier. The explanation model is g: $R\ d\ 0 \rightarrow R$, $g \in G$, where G is a class of potentially interpretable models, such as linear models, decision trees, or rule lists; given a model $g \in G$, it can be visualized as an explanation to the human expert. (For details please refer to Ribeiro et al. [12].)

14.5.2 Ante-Hoc Systems

Ante-hoc systems are interpretable by design towards glass-box approaches [13]; typical examples include linear regression, decision trees and fuzzy inference systems. The latter have a long tradition and can be designed from expert knowledge or from data and provide—from the viewpoint of Human computer interaction—a good framework for the interaction between human expert knowledge and hidden knowledge in the data [14].

14.6 Examples of AI-Powered IoT-Based Disease Diagnosis and Prevention Platforms

IoT-based healthcare services are usually built upon a wireless interface and connected with the internet. Hence, they support mobility. All the healthcare-related information, such as therapy, diagnosis, recovery, inventory and medication, can be collected, managed and shared efficiently by using IoT-based systems with global connectivity [15]. A recent survey showed the studies related to the use of IoT in 13 medical subfields (Figure 14.5) [16].

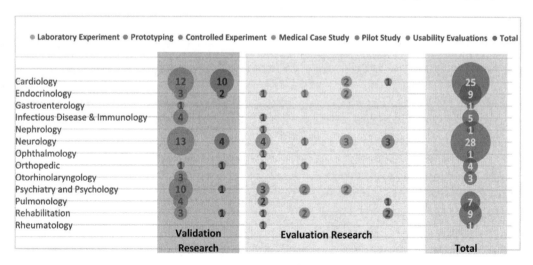

FIGURE 14.5
Bubble plot of IoT applications in medical subfields per year. (Sadoughi et al. [16]).

14.7 Explainable AI for Adverse Childhood Experiences

Brenas et al. developed an "explainable" AI system known as the Semantic Platform for Adverse Childhood Experiences Surveillance (SPACES) [17]. SPACES is an intelligent recommendation system that employs ML techniques to help in screening patients and allocating or discovering relevant resources. According to the researchers, the proposed system intends to build rapport with patients by generating personalized questions during interviews while minimizing the amount of information that needs to be collected directly from the patient.

14.7.1 Detecting Dementia

Onoda et al. developed a new screening test for dementia. The test runs on an iPad and can be utilized for mass screening, known as the Cognitive Assessment for Dementia, iPad version (CADi). The CADi consists of items involving immediate recognition memory for three words, semantic memory, categorization of six objects, subtraction, backward repetition of digits, cube rotation, pyramid rotation, trail making A, trail making B and delayed recognition memory for three words [18].

Cognetivity Neurosciences developed a sensitive diagnostic tool that is able to perform a full cognitive assessment test on Apple iOS devices [19]. During this 5-minute test, several natural images are briefly shown to participants, who are asked to respond as quickly and accurately as possible to indicate whether they have seen a pre-specified image category. Their answers are then compared with a dataset by the AI algorithms that will cluster test performance in terms of accuracy, speed and image properties. Depending on the answers, in fact, the performance of large areas of the brain can be evaluated, detecting the earliest signs of dementia and other mental impairment, which can also be monitored for progression remotely.

14.7.2 Diagnosing Autism with an HIPAA-Compliant Platform

Cognoa, a digital behavioral health company, developed an AI-based diagnostic tool that is able to spot early signs of autism in children. With a 90% accuracy rate, this software is so precise and efficient that it has already received Breakthrough Device Designation from the U.S. Food and Drug Administration (FDA), giving it priority review. This tool has not only lowered behavioral healthcare costs but it has also ensured the anonymous collection of all sensitive data used to feed these algorithms [20,21]. Cognoa hired Immuta, a company that developed a software platform that could help enforce data access roles, permissions and policies, which would meet all compliance requirements. Immuta's platform expedites the process of collecting data required by AI and allows for the delivery of Health Insurance Portability and Accountability Act (HIPAA) compliant algorithms within the most stringent of regulatory environments.

14.7.3 Preventing Heart Diseases

Potentially lethal heart diseases such as aortic stenosis (AS) can only be detected by highly skilled physicians. Minuscule variations in heartbeat, such as a heart murmur, must be heard and identified with a stethoscope, and it is not infrequent that these sounds are so faint and subtle that they end up being misdiagnosed.

Smart algorithms such as the one used by Eko's AI are able to detect the sounds associated with AS with a 97.2% precision rate. Other AI-based technologies used to detect heart issues are the wearable Zio Patch developed by iRhythm Technologies that monitors heart rate activity for up to 2 weeks, and Apple's ECG app, which can detect early signs of atrial fibrillation [22].

Very recently, Dave et al. described the examples based on the heart disease dataset and described how the explainability techniques should be preferred to create trustworthiness while using AI systems in healthcare [23].

14.7.4 Preventing Pressure Injuries

Pressure injuries (often called bedsores or decubitus ulcers) are a serious issue for patients with limited mobility or who must spend most of their time in a chair or bed. Some of these injuries never heal completely, even after treatment, and in the United States, they cost up to $11.6 billion every year.

ML-based models have been developed to correctly predict and assess risk for pressure injuries in critical care patients by examining large amounts of clinical data readily available in the patient records. These models can identify those patients who will benefit most from specialty interventions that are too expensive to be used for every patient [24].

14.7.5 AI-Powered Pediatric Services

Enable My Child is a pediatric therapy platform that provides online, in-person and hybrid speech, occupational, psychological and physical therapy to children across the United States. Their AI-based platform, called The EMC Brain, uses machine learning algorithms to track specific intervention strategies that have been successful with students of various ages, abilities and conditions and uses that data to help formulate effective treatment plans [25].

Therapists can access The EMC Brain for evidence-based data to assist with therapy planning to make therapy more effective and yield better results for students. The EMC Brain serves as a vehicle for delivering therapy and utilizes AI to collect data on treatment plans and make recommendations as the system grows smarter.

14.8 Challenges of XAI in IoT-Based Platform and Future Prospects

At present, XAI in IoT is in its nascent research field. Deep learning approaches of ML are helping in the explainability. However, it is hoped that sufficient progress in this field will lead to both power and accuracy as well as the required transparency and explainability. Actions of AI should be traceable to a certain level that can be determined by the consequences that can arise from the AI system. Important AI systems should have a significant explanation and transparency requirements to prevent further errors. However, levels of transparency or explainability may vary system to system. For example, product recommendation systems may allow lower levels of transparency, but medical diagnosis systems require higher levels of explainability and transparency. There are efforts through standards organizations to arrive at common, standard understandings of these levels of transparency to facilitate communication between end users and technology vendors [5].

To maintain the quality of AI system associated with an IoT device, a standard policy is crucial. Organizations need to have governance over the operation of their AI systems. Oversight can be achieved through the creation of committees or bodies to regulate the use of AI. These bodies will oversee AI explanation models to prevent roll-out of incorrect systems. Such an agency is the U.S. Defense Advanced Research Project Agency (DARPA). DAPRA is trying to produce explainable AI solutions through a number of funded research initiatives [26]. DARPA describes AI explainability in three parts: (i) prediction accuracy, which means models will explain how conclusions are reached to improve future decision making, (ii) decision understanding and trust from human users and operators, as well as (iii) inspection and traceability of actions undertaken by the AI systems. Traceability will enable humans to get into AI decision loops and have the ability to stop or control the tasks whenever the need arises. An AI system is not only expected to perform a certain task or impose decisions but also have a model with the ability to give a transparent report of why it made specific conclusions.

In the healthcare sector, where mistakes can have catastrophic effects, the black box aspect of AI makes it difficult for doctors and regulators to trust it. Doctors are trained primarily to identify the outliers, or the strange cases that do not require standard treatments. If an AI algorithm is not trained properly with the appropriate data, and we can not understand how it makes its choices, we can not be sure it will identify those outliers or otherwise properly diagnose patients, for instance. For these same reasons, the black box aspect of AI is also problematic for the FDA, which currently validates AI algorithms by looking at what type of data is input into the algorithms to make their decisions on the data. Furthermore, many AI-related innovations pass through the FDA because a doctor stands between the answer and the final diagnosis or an action plan for the patient. For example, in its latest draft guidance (released on Sept. 28, 2020) the FDA continues to require doctors to be able to independently verify the basis for the software's recommendations in order to avoid triggering higher scrutiny as a medical "device." Thus, software is lightly regulated where doctors can validate the algorithm's answers. Consider the case of a medical image, where doctors can double-check suspicious masses highlighted by the algorithm. With algorithms such as deep learning, however, the challenge for physicians is that they have no context for why a diagnosis was chosen [5].

Explainable AI calls for confidence, safety, security, privacy, ethics, fairness and trust [27] and puts usability on the research agenda, too [28]. All these aspects together are crucial for applicability in medicine generally, and for future personalized medicine specifically [29].

Recently, Ellahham et al. reviewed AI in healthcare along with its implication for safety (Table 14.1). Strategies for safety of AI and ML in healthcare are evolving and are not yet fully developed. The cost and distribution of outcomes in AI-based systems are not precisely known. Large feasibility studies and cost-effectiveness assessments can help improve adoption of AI in health care. Privacy, sharing and disclosure of safety data relating to AI applications should be strengthened. High standards should be defined for validation of AI and ML applications in health care. Methods, guidelines and protocols should be formulated to enable the safe and effective development and adoption of AI and ML in healthcare. Trust and training will allow the full functional integration of AI into research and practice in healthcare [1].

TABLE 14.1

Safety Issues for Artificial Intelligence (AI) in Healthcare (Ellahham et al. [1])

Safety Issue	Elements of Hazard	Key Steps to Mitigation
Distributional shift	Out-of-sample predictions	Training of AI systems with large and diverse datasets
Quality of datasets	Poor definition of outcomes Non-representative datasets	Build more inclusive training algorithms using balanced datasets, correctly labeled for outcomes of interest
Oblivious impact	High rates of false-positive and false-negative outcomes	Include outliers in training datasets Enable systems to adjust for confidence levels
Confidence of prediction	Uncertainty of predictions Automation complacency	Sustained and repeated use of AI algorithms Transparent and easily accessible AI algorithms
Unexpected behaviors	Calibration drifts	Design and train systems to learn and unlearn and have more predictable behaviors
Privacy and anonymity	Identification of patient data	Define layers of security and rules for data privacy Anonymize data before sharing
Ethics and regulations	Poor ethical standards and regulatory control for development and deployment of AI	Build regulatory reforms to support integration of AI in healthcare

14.9 "Omics" Data, Bioinformatics and Future Scope of XAI in Disease Diagnostics

Systems biology, which includes different omics technologies, is able to explain cellular mechanisms. Modern biology harnesses its power from technological advances in the field of -omics and the advent of next-generation sequencing (NGS). These technologies provide a spectrum of information ranging from genomics, transcriptomics and proteomics to epigenomics, pharmacogenomics, metagenomics and metabolomics. The large volume of data generated by these high-throughput methodologies necessitates the need to analyze, integrate and concurrently interpret this avalanche of information in a systemic way [30].

However, currently, platforms are missing that help towards not only the analysis but the interpretation of the data, information and knowledge obtained from the above-mentioned -omics technologies and the ability to cross-link them to hospital data. Moreover, it is necessary to narrow the gap between genotype and phenotype as well as provide additional information regarding biomarker research and drug discovery, where biobanks [31] play an increasingly important role. One of the grand challenges here is to close the research cycle in such a way that all the data generated by one research study can be consistently associated with the original samples; therefore, the underlying original research data and the knowledge gained thereof can be reused. This can be enabled by a catalogue providing the information hub connecting all relevant information sources [32]. The key knowledge embedded in such a biobank catalogue is the availability and quality of proper samples to perform a research project. To overview and compare collections from different catalogues, visual analytics techniques are necessary, especially glyphs-based visualization techniques [33]. We cannot emphasize often enough the combined view on hetero-geneous data sources in a unified and meaningful way, consequently enabling the discovery and visualization of data from different sources, which would enable to-tally new insights. Here, toolsets are urgently needed to support the bidirectional interaction with computational multi-scale analysis and modeling to help to go to-ward the far-off goal of future medicine [34,35].

14.10 Legal and Ethical Concerns

In the past few years, AI-assisted data analysis and learning tools have been used widely in research with patient electronic health records (EHRs). These records were previously kept on paper within hospitals but now exist electronically on secure computer servers. The increasing computational power of hardware and AI algo-rithms is also enabling the introduction of platforms that can link EHRs with other sources of data, such as biomedical research databases, genome sequencing data-banks, pathology laboratories, insurance claims and pharmacovigilance surveillance systems, as well as data collected from mobile IoT devices such as heart rate monitors. AI-assisted analysis of this big data can generate clinically relevant information in real time for health professionals, health systems administrators and policymakers. These clinical decision support systems (CDSS) are programmed with rule-based systems, fuzzy logic, artificial neural networks, Bayesian networks, as well as general machine-learning techniques [36]. CDSS with learning algorithms are currently under devel-opment to assist clinicians with their decision making based on prior successful di-agnoses, treatment, and prognostication. However, the nature of machine learning means that the large datasets on which the tools are trained may not be made available to the public for scrutiny. By its very nature, AI-driven systems are unique and new in how they function when compared with other types of CDSS that may have been in existence for decades; hence, they present issues more acute than those of previous types of CDSS [37]. The key issues related to AI-driven healthcare are: (i) account-ability and transparency of the decisions AI-assisted systems make, (ii) the potential for group harm arising from biases built into AI algorithms, (iii) public interest in generating more efficient healthcare from AI-assisted systems, and (iv) the potential conflicting professional roles and duties of clinicians as both users and generators of research data for AI-assisted systems. Besides, various technicalities during the decision-making process of AI give rise to concerns that might have ethical and legal import. For example, biased training data, inconclusive correlations, intelligibility, inaccuracy and discriminatory outcomes [38].

14.11 Conclusion

This chapter elaborated on the role of XAI in IoT-based disease prediction and diagnosis. The content was divided into five sections such that the first section discussed on the basic concepts on the working of XAI followed by the basic principles behind the application of XAI in IoT-based disease prediction and diagnosis. The second section focused on the XAI models applied in disease prediction and diagnosis. The third section addressed the healthcare use cases based on IoT. The fourth section discussed the challenges and prospects in this field. The fifth section elaborated on the legal and ethical associations of XAI in disease diagnosis.

References

1. Ellahham S., Ellahham N., Simsekler M. C. (2020). "Application of artificial intelligence in the health care safety context: Opportunities and challenges." *American Journal of Medical Quality.* vol. 35, no. 4, pp. 341–348.
2. Atziori L., Iera A., Morabito G. (2010). "The internet of things: A survey." *Computer Networks.* vol. 54, no. 15, pp. 2787–2805
3. Aktas F., Ceken C., Erdemli Y. E. (2018). "IoT-based healthcare framework for biomedical applications." *Journal of Medical and Biological Engineering.* vol. 38, no. 6, pp. 966–979.
4. Al Turjman F., Baali I. (2019). "Machine learning for wearable IoT-based applications: A survey." *Transactions on Emerging Telecommunications Technologies.* p. e3635.
5. Schmelzer, R. (2019) "Making The Internet Of Things (IoT) More Intelligent With AI." www.forbes.com.
6. Hoffman R. R., Johnson M., Bradshaw J. M., Underbrink A. (2013). "Trust in Automation." *IEEE IEEE Intelligent System.* vol. 28, no. 1, pp. 84–88.
7. Asan O., Bayrak A. E., Choudhury A. (2020). "Artificial intelligence and human trust in healthcare: Focus on clinicians." *Journal of Medical Internet Research.* vol. 22, no. 6, pp. e15154.
8. Xu F., Uszkoreit H., Du Y., Fan W., Zhao D., Zhu J. (2019). "Explainable AI: A brief survey on history, research areas, approaches and challenges." In *CCF International Conference on Natural Language Processing and Chinese Computing* (pp. 563–574). Springer, Cham.
9. Niv Mizrahi (2020), "Explainable AI: The key to responsibly adopting AI in medicine." https://insidebigdata.com/2020/04/16/explainable-ai-the-key-to-responsibly-adopting-ai-in-medicine/ April 16, 2020
10. Holzinger A., Langs G., Denk H., Zatloukal K., Müller H. (2019). "Causability and explainability of artificial intelligence in medicine." *Wiley Interdisciplinary Reviews: Data Mining and Knowledge Discovery.* vol. 9, no. 4, pp. e1312.
11. Ribeiro M. T., Singh S., Guestrin C. (2016). ""Why should I trust you?" Explaining the predictions of any classifier." In Proceedings of the *22nd ACM SIGKDD international conference on knowledge discovery and data mining* (pp. 1135–1144).
12. Ribeiro M. T., Singh S., Guestrin C. (2016). "Model-agnostic interpretability of machine learning." arXiv preprint arXiv:1606.05386.
13. Holzinger A., Biemann C., Pattichis C. S., Kell D. B. (2017). "What do we need to build explainable AI systems for the medical domain?." arXiv preprint arXiv:1712.09923.
14. Guillaume S. (2001). "Designing fuzzy inference systems from data: An interpretability-oriented review." *IEEE Transactions on Fuzzy Systems.* vol. 9, no. 3, pp. 426–443.
15. Pang Z., Tian, J. (2013). "Ecosystem analysis in the design of open platform based in-home healthcare terminals towards the internet-of-things." In *15th International conference on advanced communication technology* (pp. 529–534).
16. Sadoughi F., Behmanesh A., Sayfouri N. (2020). "Internet of things in medicine: A systematic mapping study." *Journal of Biomedical Informatics.* vol. 103, p. 103383.

17. Brenas J. H., Shin E. K., Shaban-Nejad A. (2019). "Adverse childhood experiences ontology for mental health surveillance, research, and evaluation: Advanced knowledge representation and semantic web techniques." *JMIR Mental Health.* vol. 6, no. 5, pp. e13498.

18. Onoda K., Hamano T., Nabika Y., Aoyama A., Takayoshi H., Nakagawa T., Ishihara M., Mitaki S., Yamaguchi T., Oguro H., Shiwaku K. (2013). "Validation of a new mass screening tool for cognitive impairment: Cognitive Assessment for Dementia, iPad version." *Clinical Interventions in Aging.* vol. 8, pp. 353.

19. Owens A. P., Ballard C., Beigi M., Kalafatis C., Brooker H., Lavelle G., Brønnick K. K., Sauer J., Boddington S., Velayudhan L., Aarsland D. (2020). "Implementing remote memory clinics to enhance clinical care during and after COVID-19." *Frontiers in Psychiatry.* vol. 11, pp. 990.

20. Tariq Q., Daniels J., Schwartz J. N., Washington P., Kalantarian H., Wall D. P. (2018). "Mobile detection of autism through machine learning on home video: A development and prospective validation study." *PLoS Medicine.* vol. 15, no. 11, pp. e1002705.

21. Washington P., Park N., Srivastava P., Voss C., Kline A., Varma M., Tariq Q., Kalantarian H., Schwartz J., Patnaik R., Chrisman B. (2019). "Data-driven diagnostics and the potential of mobile artificial intelligence for digital therapeutic phenotyping in computational psychiatry." *Biological Psychiatry: Cognitive Neuroscience and Neuroimaging.* vol. 5, no. 8, pp. 759–769.

22. Claudio Buttice (2020). "Top 20 AI use cases: Artificial intelligence in healthcare." https://www.techopedia.com/top-20-ai-use-cases-artificial-intelligence-in-healthcare/2/34 047. March 2, 2020

23. Dave D., Naik H., Singhal S., Patel P. (2020). "Explainable AI meets healthcare: A study on heart disease dataset." arXiv preprint arXiv:2011.03195.

24. Malvade P. S., Joshi A. K., Madhe S. P. (2017). "IoT based monitoring of foot pressure using FSR sensor." In *2017 International Conference on Communication and Signal Processing (ICCSP)* (pp. 0635–0639). IEEE.

25. Nugraha A., Izah N., Hidayah S. N., Zulfiana E., Qudriani M. (2019). "The effect of gadget on speech development of toddlers." *Journal of Physics: Conference Series.* vol. 1175, no. 1, p. 012203. IOP Publishing.

26. Challenge D. R. (2019). "Defense advanced research projects agency."

27. Kieseberg P., Weippl E., Holzinger A. (2016). "Trust for the doctor-in-the-loop." *ERCIM News.* vol. 104, no. 1, pp. 32–33.

28. Miller T., Howe P., Sonenberg L. (2017). "Explainable AI: Beware of inmates running the asylum or: How I learnt to stop worrying and love the social and behavioural sciences." arXiv preprint arXiv:1712.00547.

29. Hamburg M. A., Collins F. S. (2010). "The path to personalized medicine." *New England Journal of Medicine.* vol. 363, no. 4, pp. 301–304.

30. Holzinger A., Plass M., Holzinger K., Crisan G. C., Pintea C. M., Palade V. (2017). "A glass-box interactive machine learning approach for solving NP-hard problems with the human-in-the-loop." arXiv preprint arXiv:1708.01104. 2017 Aug 3.

31. Huppertz B., Holzinger A. (2014). "Biobanks–a source of large biological data sets: open problems and future challenges." In *Interactive Knowledge Discovery and Data Mining in Biomedical Informatics* (pp. 317–330). Springer, Berlin, Heidelberg.

32. Müller H., Reihs R., Zatloukal K., Jeanquartier F., Merino-Martinez R., van Enckevort D., Swertz M. A., Holzinger A. (2015). "State-of-the-art and future challenges in the integration of biobank catalogues." *InSmart Health* (pp. 261–273). Springer, Cham.

33. Müller H., Reihs R., Zatloukal K., Holzinger A. (2014). "Analysis of biomedical data with multilevel glyphs." *BMC Bioinformatics.* vol. 15, no. 6, pp. 1–2.

34. Hood L., Friend S. H. (2011). "Predictive, personalized, preventive, participatory (P4) cancer medicine." *Nature Reviews Clinical Oncology.* vol. 8, no. 3, pp. 184–187.

35. Tian Q., Price N. D., Hood L. (2012). "Systems cancer medicine: towards realization of predictive, preventive, personalized and participatory (P4) medicine." *Journal Of Internal Medicine*. vol. 271, no. 2, pp. 111–121.

36. Wagholikar K. B., Sundararajan V., Deshpande A. W. (2012). "Modeling paradigms for medical diagnostic decision support: A survey and future directions."*Journal of Medical Systems*. vol. 36, no. 5, pp. 3029–3049.

37. Lysaght T., Lim H. Y., Xafis V., Ngiam K. Y. (2019). "AI-assisted decision-making in healthcare." *Asian Bioethics Review*. vol. 11, no. 3, pp. 299–314.

38. Mittelstadt B. D., Allo P., Taddeo M., Wachter S., Floridi L. (2016). "The ethics of algorithms: Mapping the debate." *Big Data and Society*. vol. 3, no. 2. https://doi.org/10.1177/2053951716679679

15

An Overview on Sleep, Sleep Stages and Sleep Behavior

Santosh Kumar Satapathy and D. Loganathan

Department of Computer Science and Engineering, Pondicherry Engineering College, Puducherry, India

CONTENTS

DOI: 10.1201/9780429318078-15

15.1 Introduction

Sleep is a fundamental requirement for proper balance between mental and physical health. Each night, the human body needs a definite amount of sleep to function effectively. In general, the average amount of sleep needed for different age groups is not the same. The National Sleep Foundation published a sleep requirement chart for different age groups. The sleep requirement statistics are presented in Table 15.1. When a person sleeps fewer hours during the night, this reduced state is called "sleep debt", which can accumulate toward a sleep disorder. "Sleep disorder" is the term given to disturbance in sleeping patterns of persons in terms of quality and quantity. Due to disturbed sleep, the person develops an imbalance between mental and physical functioning, which further worsens other health conditions. Currently, sleep-related disorders are one of the major reasons for death in many developed countries. Not only India but the whole world is facing increased rates of sleep disorders. Many people do not realize that sleep disorders are a major cause of death. Also, people are very careless with their lifestyle and eating habits, which can lead to a sleep disorder and ultimately death. Computer science is growing rapidly, creating new fields of research and progress interfaces to tackle the disastrous situations that can help ordinary people diagnose themselves with a sleep disorder and learn more about the deadly sleep disorders with which they are suffering.

As per the survey [1], 91% of the population in India experiences sleep-related issues; however, just 3% of Indians talk about their sleep-related issues with physicians. Sleep disorders constitute a global epidemic. Along with this, sleep disorders have a significant economic impact. The economic, social and personal consequences that people with sleep disorders face are much higher than the cost of treatment. According to a report by the University of Maryland Medical Center, the annual cost of direct sleep-related problems in the United States is $16 billion, whereas indirect costs reach $50–$100 billion, which includes accidents, litigation, property destruction, hospitalization and death. Various impacts occurred due to various sleep disorders, and their effects may be short or long term. Short-term impacts include impaired attention and concentration, reduced quality of life and increased rate of absenteeism with reduced productivity and accidents at work, at home or on the road. The long-term impacts of sleep disorders include high blood pressure, obesity, cardiovascular attack, coronary artery heart failure and memory impairment as well as depression.

TABLE 15.1

Sleep Needs over the Life Cycle

Newborns/Infants	0–2 months	10.5–18 hours
	2–12 months	14–15 hours
Toddlers/Children	12–18 months	13–15 hours
	18 months to 3 yrs	12–14 hours
	3–5 yrs	11–13 hours
	5–12 yrs	10–11 hours
Adolescents	On average	9.25 hours
Adults/Older Persons	On average	7–9 hours

It is important to have proper sleep during the night; poor sleep affects quality of life. Sometimes its impact is reflected in physiological activities such as learning ability, ability to participate in physical activity, mental ability and performance of overall activities [2]. In general, complete sleep duration should cover all sleep stages at regular time intervals related to the brain's neuro system [3]. The modern digital generation has complicated the lifestyle of humans and ultimately resulted in millions of people getting poor quality sleep during the night. This problem is seen across the world with all age groups, and this is a global challenge in the healthcare sector because research has found that poor sleep is responsible for critical diseases such as bruxism [4], insomnia [5], narcolepsy [6], obstructive sleep apnea [7] and rapid eye movement behavioral disorder [8]. Finally, poor sleep damages various parts of the body, leading to heart failure, brain stroke and several neurological disorders.

Normally, sleep behavior is characterized by changes in respiration and heart rate, brain behavior and muscle movements. Currently, there are two sleep stages used during sleep staging analysis, non-rapid eye movement (N-REM) and rapid eye movement (REM), with the third stage being wakefulness (W). The first sleep handbook was edited by Rechtschaffen and Kales (R&K) in 1968 [9]. R&K further divided the N-REM sleep stage into four sub-sleep stages: S1, S2, S3 and S4. Clinicians used these sleep standards during analysis of sleep irregularities until 2008, when the American Academy of Sleep Medicine (AASM) [10] published a new sleep handbook with small modifications to the R&K rules. AASM further segmented N-REM into three sub-sleep stages: N-REM1, N-REM2 and N-REM3 [10]. The sleep cycle generally repeats at regular intervals between NREM stages to REM stages, and each sleep cycle is around 90–110 minutes [11]. The quality of sleep ratio differs from person to person according to their age. Stage 1 of N-REM sleep (N-REM1) is light, where the subject's eye movements and muscle movements are slow. During the N-REM2 sleep stage, eye movements completely stop, and brain response becomes slower. Both N-REM1 and N-REM2 sleep stages are categorized as light sleep stages. Similarly, the N-REM3 and N-REM4 stages are called deep sleep, in which no eye movements occur and some muscle movements appear [12]. Finally, in the REM stage breathing increases, and physical movements are rapid.

Sleep staging is normally examined using polysomnographic (PSG) recordings from an admitted subject in a clinic. During the PSG test, different physiological recordings are collected from the subject to measure sleep quality during the night. The major recordings include electroencephalogram (EEG), electromyogram (EMG), electrooculogram (EOG) and electrocardiogram (ECG) [13]. Among these, the first priority is the EEG signal because it provides information about the brain activity and behavior of the subject during sleep. It can help to characterize the behavior during the different sleep stages by measuring different frequency ranges of EEG signals, which are segmented into the delta band (<3.5 Hz), theta band (4–7 Hz), alpha band (8–13 Hz), beta band (14–30 Hz) and gamma band (30–80 Hz). Ultimately, it can demonstrate the characteristics of different stages of sleep [14].

It is very difficult for sleep experts to manually monitor EEG signals because they may insert errors during the 7–8 hrs of the sleep study. Sleep experts monitor EEG signals using a 30 s framework and attach the labels for the sleep stages. This approach consumes a lot of time and requires a lot of man power for the many hours of sleep recordings. To overcome this difficulty, automated sleep stage classification is used to analyze sleep recordings in real time. This step is important in designing classification on stages of sleep. Currently, overnight sleep study through polysomnography is one of

TABLE 15.2

Sleep Stages Terminology

Sleep Stages	R&K	AASM
Wake	Stage W	Stage W
NREM	Stage N1	Stage N1
	Stage N2	Stage N2
	Stage N3	Stage N3
	Stage N4	
REM	Stage REM	Stage R

AASM = American Academy of Sleep Medicine; NREM = non-rapid eye movement; R&K = Rechtschaffen and Kales A; REM = rapid eye movement; stages N3 and N4 are combined into stage N3.

the standard procedures for measuring sleep irregularities [15]. Some sleep disorders are serious and greatly affect the physical, psychological, cognitive and motor functioning of a person [16,17]. The AASM estimates that 40 million Americans suffer from chronic sleep disorders [1,18].

The R&K sleep manual was revised after the publication of the AASM manual. The NREM sleep stages are now only three stages, N1, N2 and N3 (N3 and N4 were combined as one stage). The old and new rules for sleep stages are shown in Table 15.2.

15.2 10–20 Electrode Placement System of EEG Recording

The International Federation of Societies for Electroencephalography recommended the 10–20 electrode placement system for recording brain signals through placement of multiple electrodes on the human scalp. The same concept also was approved by the clinical community [19]. According to this placement standard, the head is segmented from specific skull landmarks such as the nasion, preauricular and inion to adequately cover all regions of the brain, as shown in Figure 15.1. The labeling of the electrode placement 10–20 indicates that the electrodes are placed in the central midline at 10%, 20%, 20%, 20%, 20% and 10% of the total nasion-inion distance; the other electrodes are placed in the same way at fractional distances [20].

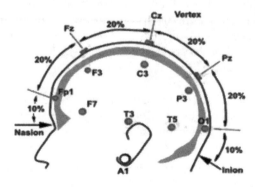

FIGURE 15.1
10–20 electrode placement system.

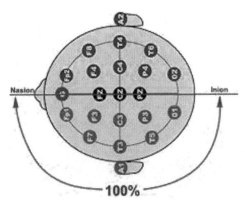

FIGURE 15.2
Notations for placement of electrodes on the scalp.

15.2.1 Electrodes

Electrodes are generally used to record the brain behaviors of subjects. They are placed on the surface of the scalp. Different types of electrodes are used, such as pre-galled or galled; some electrodes are designed as disc styles and are made of gold, silver or stainless steel. Other forms of electrodes are headbands and electrode caps as well as saline-based and needle electrodes. Two different ways the EEG signals are recorded is unipolar recording and bipolar recording. During unipolar recording, only one reference electrode is used in reference to the other active electrodes. Similarly, in the bipolar recording system, an extra electrode is placed for tracking the difference in voltage between two locations. The standard 10–20 electrode placement and placement of electrode notations are presented in Figure 15.1 and Figure 15.2.

15.2.2 Montage

Montage is used for the purpose of multiple derivations for a multichannel recording system. The recording configuration can be 64, 128 or 256. This recording procedure was accepted internationally; the different notations are followed during placement of electrodes on the scalp as follows: Frontal pole (FP), Frontal (F), Central (C), Parietal (P), Occipital (O) and Temporal (T). The detailed description of the channels represents different segments of the brain, presented in Table 15.3. The notations are managed with

TABLE 15.3

Electrode and Channel Information

Brain Region	Left	Midline	Right
Frontal Pole	Fp1	Fpz	Fp2
Frontal	F3	Fz	F4
Inferior Frontal	F7		F8
Anterior Temporal	F9 or A1		F10 or A2
Mid Temporal	T3		T4
Posterior Temporal	T5		T6
Central	C3	Cz	C4
Parietal	P3	Pz	P4
Occipital	O1	(Oz)	O2

(a)

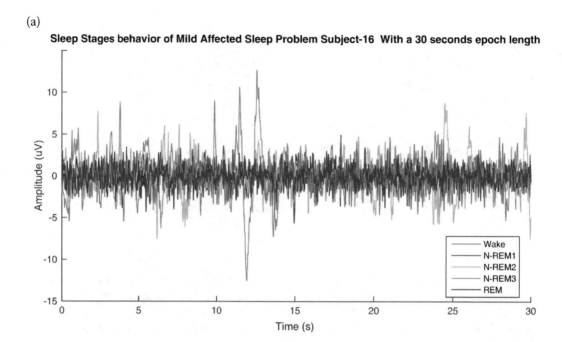

(b)

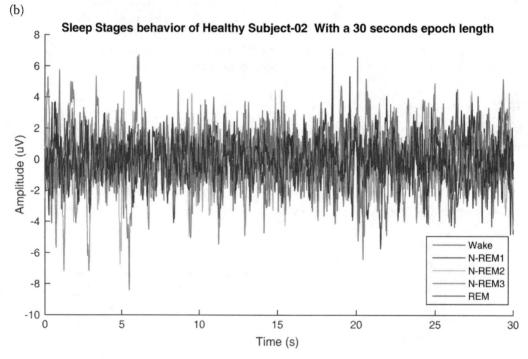

FIGURE 15.3

The brain behavior recorded using EEG signals: (a) subject with a sleep disorder, (b) healthy control subject. The pictorial representation of the sleep life cycle are presented in Figure 15.18 [22].

odd and even numbers. Odd numbers are located on the left hemisphere, and even numbers are located on the right hemisphere. Symbol A is the ear lobe for reference. The most popular multi-channel recordings are done through electrode caps. The different channels used for monitoring the brain behavior from different regions from the brain are presented in the Figure 15.3 [21].

15.2.2.1 Electroencephalogram (EEG) Electrodes

C3-A2, C4-A1, O1-A2, O2-A1, left central, right central, left occipital, right occipital. Electrode location: ground (FPZ), reference (CZ), A1 or M1 and A2 or M2; C3 and O1, left central and occipital, respectively, EEG electrodes; C4 and O2, right central and occipital, respectively, EEG electrodes. The exploring electrodes, such as F3, F4, C3, C4, O1 and O2, are selected from the opposite side of the head from the mastoid electrode (M1, M2) or average (AVG).

15.2.2.2 Electrooculogram (EOG) Electrodes

Right outer canthus: ROC-A1; left outer canthus: LOC-A2; LOC-A2, ROC-A1; left and right electrooculogram referred to right and left mastoid leads; M2, right mastoid electrode location; M1, left mastoid electrode location.

15.2.2.3 Electromyogram Electrodes (EMG)

Left tibialis anterior (LtTib1-LtTib2) and right tibialis anterior (RtTib1-RtTib2) EMG electrodes; Chin1-Chin2, submental EMG signal, limb EMG (left leg, right leg).

15.2.2.4 Electrocardiogram Electrodes (ECG)

ECG1-ECG2, ECG2-ECG3.

15.2.2.5 Respiratory Electrodes

SNORE; THOR/CHEST and ABD; THOR1-THOR2 (thoracic effort channel); ABD1-ABD2 (abdominal effort channel); CPAP (continuous positive airway pressure).

15.3 Human Sleep Staging Parameters

During PSG monitoring, three physiological parameters are commonly used (Figure 15.4):

- EEG leads: one occipital, one central and one frontal lead
- EOG leads: left eye and right eye
- EMG leads: one chin EMG (submental)

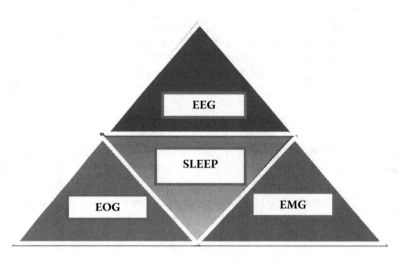

FIGURE 15.4
Scoring of sleep stages.

15.3.1 Electroencephalographic Recording

EEG recording during the PSG study is used for interpreting the patterns of the scalp EEG. The EEG records electrical potentials generated by the motor cortex but can reflect the influence of deeper structures of the brain, such as the thalamus. The EEG signal recording is done through different fixed electrodes placed on the scalp. Measurement of the EEG signal through different channels varies based on electrode position on the scalp. The PSG references the left or right electrodes to the electrodes on the opposite right and left ears (A2, A1) or mastoids (M2, M1). The 30 s sleep behavior recordings from different channels of EEG signal are presented in Figures 15.5, 15.6, 15.7, 15.8, 15.9 and 15.10 for channels F3-A2, C3-A2, F4-A1, C4-A1, O1-A2, and O2-A1, respectively.

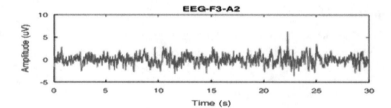

FIGURE 15.5
Left frontal region recording (typical activity of the EEG (F3-A2).

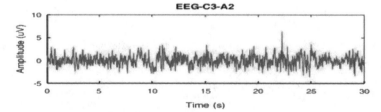

FIGURE 15.6
Left central region recording (typical activity of the EEG [C3-A2]).

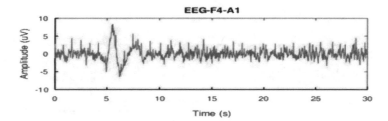

FIGURE 15.7
Right frontal recording (typical activity of the EEG [F4-A1]).

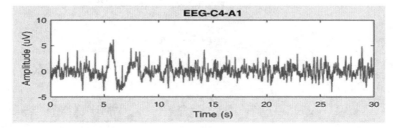

FIGURE 15.8
Right central recording (typical activity of the EEG [C4-A1]).

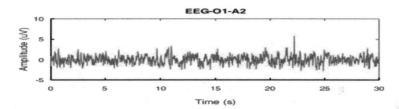

FIGURE 15.9
Left occipital recording (typical activity of the EEG [O1-A2]).

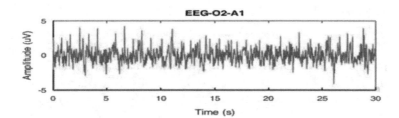

FIGURE 15.10
Right occipital region recording (typical activity of the EEG [O2-A1]).

15.3.2 Electrooculogram Recording

This phase of recording captures the changes in the electrical potential of the positive anterior aspect of the eye, the cornea, relative to the negative posterior aspect, the retina. During recording, electrodes are placed horizontally near the outer canthi or

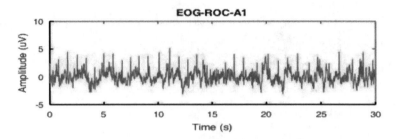

FIGURE 15.11
Right outer canthus region recording (typical activity of the EOG [ROC-A1]).

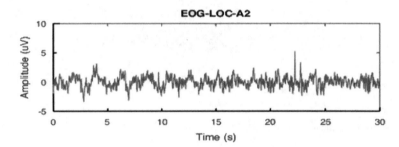

FIGURE 15.12
Left outer canthus region recording (typical activity of the EOG [LOC-A2]).

vertically 1 cm below (LOC) and 1 cm above (ROC) for capturing the transient changes of actual eye movements during sleep. Both the eye movements were monitored through two channels that are ROC-A1 and LOC-A2 from right and left eyes, respectively, and the same 30 s epoch representation are presented in Figure 15.11 and Figure 15.12.

15.3.3 Electromyography Recording

This signal provides potential information on muscle twitch, used to distinguish the sleep stages based on the EMG behavior during sleep. During REM sleep, muscle activity is minimal. Sometimes EMG channels show an intrusion of an artifact such as yawns, swallows and teeth grinding. Body movements like chin EMG (X1), right leg limb (X2) and left leg limb (X3) behavior are presented in Figure 15.13, 15.14 and 15.15, respectively.

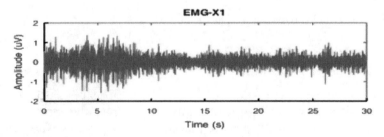

FIGURE 15.13
Chin EMG region recording (typical activity of the EMG (EMG-X1).

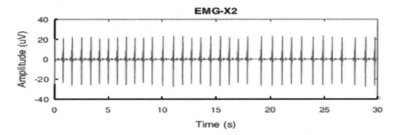

FIGURE 15.14
Limb EMG (left leg) region recording (typical activity of the EMG [EMG-X2]).

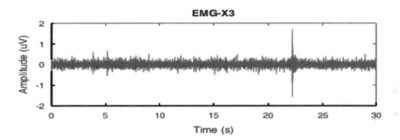

FIGURE 15.15
Limb EMG (right leg) region recording (typical activity of the EMG [EMG-X3]).

15.4 Behavior of EEG During Wakefulness and Sleep

EEG signals are recommended by sleep experts during sleep studies to identify abnormalities in different transitions of sleep states. EEG signals provide information about brain activities during sleep, which ultimately monitor abnormality disorders by observing six different types EEG waveform patterns. These wave patterns are used to differentiate between wakefulness and sleep states and to classify the sleep stages:

1. Alpha activity
2. Theta activity
3. Vertex sharp waves
4. Sleep spindles
5. K-complexes
6. Slow wave activity

The cortical behavior can be characterized by specific frequency levels. The term "frequency" is defined as the number of times a wave occurs in a specific time period (basically 1 s), and it represents cycles per second (i.e., Hertz [Hz]). In general, the EEG activity is divided into four bands based on the different frequency and amplitude levels of the waveform, and the different bands are named in terms of Greek letters (alpha, beta, theta, and delta). These bands are seen during sleep through EEG recordings with different frequency ranges. Theta is between 4 Hz and 8 Hz, alpha is in between 8 Hz and 13 Hz, beta is greater than 13 Hz, and delta is less than 4 Hz. Another

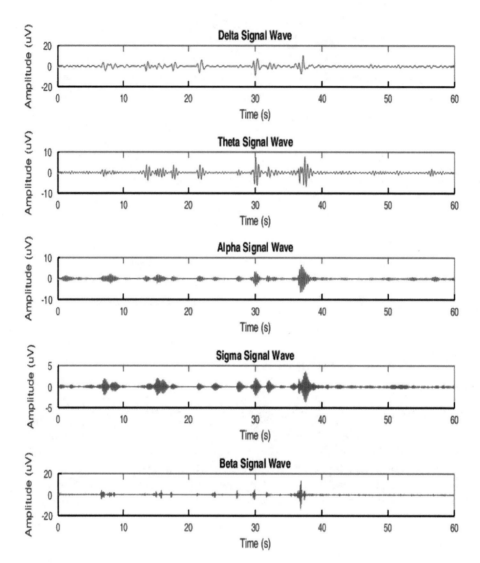

FIGURE 15.16
Sleep EEG behavior of the subject with a sleep disorder from different EEG signal sub-bands.

EEG activity that is sometimes considered during sleep analysis is gamma, whose frequency ranges from 30 Hz to 45 Hz. Sleep behavior from EEG signal sub-bands from subjects with sleep disorders and those who are healthy are presented in Figure 15.16 and Figure 15.17.

15.4.1 Alpha Activity

This activity is seen in general while the subjects are quite alert with eyes closed. Eye opening causes the alpha waves to react or decrease in amplitude. The activity appears like sinusoidal waveforms. The frequency ranges decrease with age.

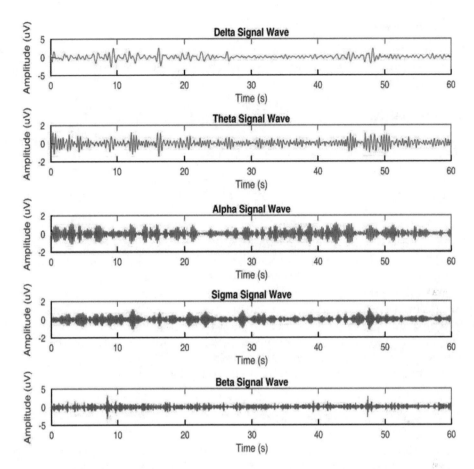

FIGURE 15.17
Sleep EEG behavior of the healthy control subject from different EEG signal sub-bands.

15.4.2 Theta Activity

In general, this activity originates in the central vertex region. For this band, there is no amplitude level, and it is the most appropriate for EEG sleep frequency.

15.4.3 Beta Activity

This activity originates in the frontal and central regions. It occurs during wakefulness and the drowsiness phases of sleep. It may be more persistent during drowsiness and may diminish during deeper sleep.

15.4.4 Sleep Spindles

Sleep spindles are seen with frequency ranges from 11 Hz to 16 Hz. They occur in the central vertex region. The duration of this wave form continues for least for 0.5 s for the purpose of sleep scoring.

15.4.5 K-Complexes

The duration of this type of waveform continues for at least 0.5 s. It is generally looks like sharp and slow waves. The waves do not have amplitude criteria. They originate in the central vertex region.

15.4.6 Delta Activity

This activity is seen in the frontal region. Delta activity has an amplitude criterion of 75 μV or greater. There is no time window for this waveform pattern.

15.5 Sleep Stages

Sleep is a naturally occurring state identified by decreasing or missing awareness of surrounding sand suspension of sensations. In this state, all muscles are in the relaxed state. It is an increased anabolic state that helps in the development of a strong immune system, nerves, muscles and skeletal system. The stages of human sleep are defined on the basis of characteristic patterns in the electroencephalogram (EEG), electrooculogram (EOG; a measure of eye movement activity) and the surface electromyogram (EMG). The continuous recording of this array of electrophysiologic parameters to define sleep and wakefulness is termed "polysomnography".

The two main types of sleep generally occur in the cyclic process:

1. Rapid eye movements (REM)
2. Non-rapid eye movements (NREM)

Throughout the night, these two phases are repeated in cyclic manner with an average duration of one cycle, typically around 45–90 min. Sleep comes in cycles of REM and NREM, frequently four or five of them for each night; the arrangement is N1 → N2 → N3 → N2 → REM. The pictorial representation of sleep life cycle are presented in Figure 15.18. A committee led by Rechtschaffen and Kales (R&K) published standardized

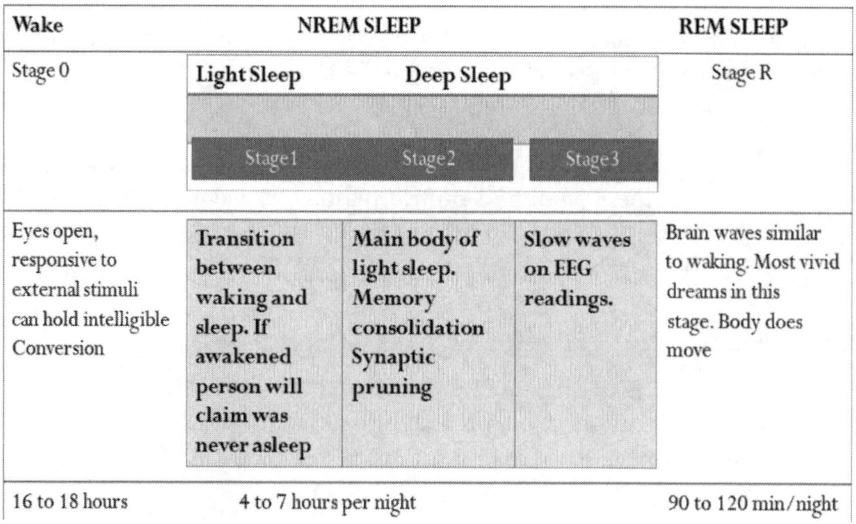

Wake	NREM SLEEP			REM SLEEP
Stage 0	Light Sleep	Deep Sleep		Stage R
	Stage 1	Stage 2	Stage 3	
Eyes open, responsive to external stimuli can hold intelligible Conversion	Transition between waking and sleep. If awakened person will claim was never asleep	Main body of light sleep. Memory consolidation Synaptic pruning	Slow waves on EEG readings.	Brain waves similar to waking. Most vivid dreams in this stage. Body does move
16 to 18 hours	4 to 7 hours per night			90 to 120 min/night

FIGURE 15.18
General structure of the sleep life cycle.

rules in relation to the process of recordings and scoring of sleep stages in 1968. According to R&K, non-REM sleep stages are further divided into four sleep sub-stages [22].

According to the *R&K Manual for the Scoring of Sleep and Associated Events*, the sleep cycle of adults consists of six sleep stages: Wakefulness, REM stage and four non-REM stages: N1, N2, N3 and N4. This standard became widely accepted in relation to sleep stage scoring, and its terminology is still used by different authors and sleep experts in their respective sleep disorder studies. According to R&K, the sleep stage scoring is done through sleep stage epochs. The duration of each epoch is either 20 s or 30 s. The duration of an epoch does not change until the completion of the whole recording. When two stages occur at one epoch, in this case the one which takes the largest portion in the epoch, that sleep stage is considered for that epoch in the sleep staging scoring system. In 2007, the AASM redefined the sleep stages and declared new rules in regard to sleep recordings and sleep stage scoring. According to the AASM manuals, the non-REM sleep stages are divided into three sleep substages, and in this scoring procedure N3 and N4 sleep stages are combined into one sleep stage as N3 [23]. As per the new rule of the AASM manuals, sleep is categorized into five sleep stages: Wakefulness, REM and three non-REM sleep stages: N1, N2 and N3. Each sleep stage is characterized by the presence (or the absence) of certain sleep stage events, and those behaviors are mostly reflected with the EEG signals. For sleep analysis, the EEG activity is typically separated into four characteristic frequency bands: delta, theta, alpha and beta.

15.5.1 NREM Stage 1

Stage 1 is the lightest stage of NREM sleep. Often defined by the presence of slow eye movements, this drowsy sleep stage can be easily disrupted, causing awakenings or arousals. Muscle tone throughout the body relaxes, and brain wave activity begins to slow. Occasionally people may experience hypnic jerks or abrupt muscle spasms and may even experience the sensation of falling while drifting in and out of Stage 1.

In Stage 1, the brain produces high amplitude theta waves, which are very slow brain waves. This period of sleep lasts only a brief time (around 5–10 min). If you awaken someone during this stage, they might report that they were not really asleep. The sleep behavior of the subjects with reference to EEG, EOG and EMG channels are presented in Table 15.4.

TABLE 15.4

N1 Stage Scoring Rules

Stage NREM1 Rules
Subjects who generate alpha rhythm with eye closure, scored as stage NREM1 if
EEG: Alpha rhythm is attenuated and replaced by mixed-frequency activity for more than 50% of the epoch. Sometimes vertex waves also occurred, but this is not necessary for scoring the NREM1 stage.
EOG: Slow rapid eye movements happen but are not necessary for scoring stage NREM1.
EMG: Amplitude varies but is less than during the wake stage.
Some subjects who do not generate the alpha rhythm are also considered.
NREM1 stage with the following conditions
• Vertex sharp waves
• Slow eye movements
EEG frequency ranges of 4 Hz to 7 Hz activity with slowing of background frequency as 1 Hz.

TABLE 15.5

N2 Stage Scoring Rules

Stage NREM2 Rules

A. RULE DEFINING THE START OF N2 SLEEP

Subjects who generate alpha rhythm with eye closure are scored as stage NREM1 if

EEG: Considered N2 stage if one or more of the following activities occurred during either the first half of the current epoch or the last half of the previous epoch:
  a. One or more k-complexes without arousals
  b. One or more trains of sleep spindles

EOG: In general, no eye movements occurred and slow eye movements have ended.

EMG: Variable amplitude occurs but in comparison to the wake stage it is less.

B. RULE DEFINING THE CONTINUATION OF STAGE N2 SLEEP

To consider an epoch with low amplitude mixed frequency EEG activity as N2 without k-complexes or sleep spindle then it must be preceded by an epoch containing:
  a. K-complexes with unassociated arousals
  b. Sleep spindles

C. RULE DEFINING THE END OF A PERIOD OF STAGE N2 SLEEP

We can end with stage N2 if the following conditions occurred during
sleep
  a. An arousal
  b. A major body movement with rapid eye movements
  c. A transition to other stages like wake, NREM3 or REM

15.5.2 NREM Stage 2

The first actual stage of defined NREM sleep is Stage 2. In this stage, one or more K-complexes occur (not associated with an arousal) or one or more sleep spindles are seen. Eye movements cease, but the behavior of the EMG is variable and the level is less than wakefulness. The sleep characteristics during the N2 stage with reference to EEG, EMG and EOG signals are presented in Table 15.5.

15.5.3 NREM Stage 3

Generally in this stage the subject goes through deep sleep, where he or she is completely unaware of anything happening in the surrounding environment. In this stage, the body temperature and heart rhythm of the subject are very slow. The eye and muscle movements completely cease. The sleep behavior acquired from different signals such as EEG, EMG and EOG during the N3 stage are described in Table 15.6.

TABLE 15.6

N3 Stage Scoring Rules

Stage NREM3 Rules

A. An epoch can be considered as stage NREM3 if 20% or more of the epochs showed slow wave activity (SWA), irrespective of age. In the following cases, for an epoch to be scored as N3:
  a. EEG: SWA >=20% of the epochs, and sleep spindles are present.
  b. EOG: Eye movements do not typically occur during stage N3.
  c. EMG: Amplitude varies but is often lower than N2 sleep.

TABLE 15.7

REM Stage Scoring Rules

Stage REM Rules

A. For one epoch to be scored as REM, then one of the following phenomena must occur
   a. EEG: Low-amplitude with mixed frequency
   b. EOG: Rapid eye movements
   c. EMG: Low EMG tone

Continuation and End of Stage R (REM Rules B and C)

B. Continuation of REM stage: There is an absence of rapid eye movement. For an epoch to be scored as REM, then one of the following conditions must occur:
   a. EEG: Low amplitude with mixed frequency activity is present without sleep spindles and k-complexes.
   b. EMG: Tone remains slow.

C. End of Stage R: To decide to end the score of REM, then some of following conditions must occur:
   a. Subject either transitions to wake or NREM1 stage.
   b. Arousal occurs.
   c. There is an increase in chin EMG tone.
   d. Major body movements are present.
   e. Non-arousal associated sleep spindles and k-complexes are present.

15.5.4 REM Sleep

REM sleep is most commonly known as the dreaming stage. Eye movements are rapid, moving from side to side, and brain waves are more active than in Stages 2 and 3 of sleep. Awakenings and arousals occur more easily in REM; being woken during a REM period can leave one feeling groggy or overly sleepy. The sleep behavior recorded during the REM sleep phase is presented in Table 15.7.

15.6 Summary and Conclusion

Improper sleep causes serious physical and mental conditions. Sleep is important for the maintainance of day-to-day activities and has an impact on the body through secretion of hormones and management of the immune system. Improper sleep may eventually lead to sleep-related disorders. Across the globe, sleep-related diseases are increasing, and they can have a very serious impact on health. Researchers have analyzed sleep-related irregularities and their causes. From primary investigations through different surveys by health organizations and sleep laboratories, researchers have found that the major reason for sleep-related diseases is improper sleep and excessive daytime sleepiness. To diagnose sleep disorders, R&K recommended conducting a sleep study with manual techniques. Later, AASM introduced new sleep standards, and researchers developed automated sleep stage detection. Another solution proposed for diagnosing sleep disorders is proper analysis of "Sleep Stage Classification or Sleep Staging Scoring".

This chapter provided an overview of sleep, sleep stages and sleep stage scoring rules. It explored different concepts of sleep, electrophysiological parameters of sleep and changes that occur during sleep.

The major objective of this chapter is awareness about how sleep disruption affects quality of life and a person's personal and social life. Because sleep disturbances are a public health concern, we proposed this chapter on sleep and its consequences on human life. The content of this chapter is essential for all generations and for physicians who are involved in the diagnosis, treatment and management of sleep disorders.

References

1. C. Iber, S. Ancoli-Israe, A. Chesson and S. F. Quan, *For the American Academy of Sleep Medicine: The AASM Manual for the Scoring of Sleep and Associated Events: Rules, Terminology and Technical Specifications*, 1st ed. Westchester, IL: American Academy of Sleep Medicine, 2007.
2. K. A. I. Aboalayon, M. Faezipour and W. S. Almuhammadi, "Sleep stage classification using EEG signal analysis: A Comprehensive Survey and New Investigation", *Entropy*, 2016, vol. 18, pp. 272.
3. M.-H. Chung, T. B. Kuo, N. Hsu, H. Chu, K.-R. Chou and C. C. Yang, "Sleep and auto-monic nervous system changes enhanced cardiac sympathetic modulations during sleep in permanent night shift nurses," *Scandinavian Journal of Work, Environment & Health*, vol. 35, no. 3, 2009, pp. 180–187.
4. M. B. B. Heyat, F. Akhtar and S. Azad, "Comparative analysis of original wave & filtered wave of EEG signal used in the detection of bruxism medical sleep syndrome," *Int. J. Trend Sci. Res. Develop.*, vol. 1, no. 1, 2016, pp. 7–9, Nov./Dec.
5. M. B. B. Heyat, S. F. Akhtar and S. Azad, "Power spectral density are used in the investigation of insomnia neurological disorder," in *Proc. Pre Congr. Symp., Organized Indian Acad. Social Sci. (ISSA) King George Med. Univ. State Takmeelut-Tib College Hospital*, Lucknow, Uttar Pradesh, 2016, pp. 45–50.
6. F. Rahman and S. Heyat, "An overview of narcolepsy", *IARJSET*, vol. 3, 2016, pp. 85–87.
7. T. Kim, J. Kim and K. Lee "Detection of sleep disordered breathing severity using acoustic biomarker and machine learning techniques". *BioMedEngOnLine*, vol. 17, 2018, pp. 16.
8. M. Siddiqui, G. Srivastava and S. Saeed, "Diagnosis of insomnia sleep disorder using short time frequency analysis of PSD approach applied on EEG signal using channel ROC-LOC", *Sleep Science*, vol. 9, no. 3, 2016, pp. 186–191.
9. A. Rechtschaffen, "A manual for standardized terminology techniques and scoring system for sleep stages in human subjects", *Brain Information Service*, 1968.
10. C. Iber, "The AASM manual for the scoring of sleep and associated events: rules, terminology and technical specifications: American Academy of Sleep Medicine", 2007.
11. M. A. Carskadon and W. C. Dement, "Normal human sleep: An overview," in *Principles and Practice of Sleep Medicine*, M. Kryger, T. Roth, and W. C. Dement, Eds., 6th ed. Amsterdam, The Netherlands: Elsevier, 2017, pp. 15–24.
12. U. R. Acharya, O. Faust, N. Kannathal, T. Chua and S. Laxminarayan, "Nonlinear analysis of EEG signals at various sleep stages," *Computer Methods Programs Biomedicine.* vol. 80, no. 1, 2005, pp. 37–45.
13. J. V. Holland, W. C. Dement and D. M. Raynal, "Polysomnography: A response to a need for improved communication," *presented at the 14th Annu. Meeting Assoc. Psychophysiology. Study Sleep*, 1974.
14. U. R. Acharya, "Nonlinear dynamics measures for automated EEG-based sleep stage detection," *Eur. Neurol.*, vol. 74, nos. 5–6, 2015, pp. 268–287.
15. W. H. Spriggs, "Essentials of polysomnography", *World Headquarter.* MA, United States: Jones & Bartlett Pub-lishers, 2014.

16. A. Rechtschaffen and A. Kales, "A manual of standardized terminology techniques and scoring system for sleep stages of human sleep". *Los Angeles: Brain Information Ser- vice/Brain Research Institute*. UCLA, 1968.

17. The AASM manual 2007 for the scoring of sleep and associated events. *Rules, Terminology and Technical Specifications*. Westchester, IL, USA: American Academy of Sleep Medicine, 2007.

18. M. A. Caraskadon and A. Rechtschaffen, "Monitoring and staging human sleep". In *Principles and Practice of Sleep Medicine*, M. H. Kryger, T. Roth, W. C. Dement, Eds., 2nd ed. Philadelphia: WB Saunders, 2005, pp. 1359–1377.

19. P. Rau, *"Drowsy Driver Detection and Warning System for Commercial Vehicle Drivers: Field Operational Test Design, Analysis, and Progress"*, Washington, DC, USA: National Highway Traffic Safety Administration, 2005.

20. A. Ambrogetti, M. J. Hensley, and L. G. Olsen *"Sleep Disorders: A Clinical Textbook"*, London: Quay Books, 2006.

21. G. Rojas, et al. Study of resting-state functional connectivity networks using EEG electrodes position as seed. *Frontiers in Neuroscience*, vol. 12, 2018, p. 12.

22. S. K. Satapathy and D. Loganathan, "Automated Sleep Stage Classification Based on Multiple Channels of Electroencephalographic signals using Machine Learning Algorithm," *2019 International Conference On IoT Inclusive life,Springer (ICIIL 2019)*, NITTTR Chandigarh, India, 2019.

23. S. K. Satapathy, A. Mitra and D. Loganathan, "Effect of EEG Dual-Channel Acquisition and Gender Specification Subjects on the Classification of Sleep Stages using Machine Learning Techniques," *2020 3rd international conference on Emerging Technology Trends in Electronics, Communication, and Networking, Springer (ET2ECN2020), SVNIT*, Surat, India, 2020.

Index